THE MGA

"First of a new line"

Other Veloce publications -
SpeedPro Series
How to Build a V8 Engine Short Block for High Performance
by Des Hammill
How to Build & Power Tune Distributor-type Ignition Systems
by Des Hammill
How to Build & Power Tune Ford Pinto Engines
by Des Hammill
How to Build & Power Tune Weber DCOE & Dellorto DHLA Carburetors
by Des Hammill
How to Give Your MGB V-8 Power
by Roger Williams
How to Power Tune Alfa Romeo Twin Cam Engines
by Jim Kartalamakis
How to Power Tune the MG Midget & Austin-Healey Sprite
by Daniel Stapleton
How to Power Tune the MGB 4-Cylinder Engine
by Peter Burgess

General
Alfa Romeo Owner's Bible
by Pat Braden
Alfa Romeo Modello 8C 2300
by Angela Cherrett
Alfa Romeo Giulia Coupe GT & GTA
by John Tipler
Bubblecars & Microcars Colour Family Album
by Andrea & David Sparrow
Bubblecars & Microcars, More, Colour Family Album
by Andrea & David Sparrow
Bugatti 46/50 - The Big Bugattis
by Barrie Price
Bugatti 57 - The Last French Bugatti
by Barrie Price
Chrysler 300 - America's Original Muscle Car
by Robert Ackerson
Citroen 2CV Colour Family Album
by Andrea & David Sparrow
Citroen DS Colour Family Album
by Andrea & David Sparrow
Cobra - The Real Thing!
by Trevor Legate
Completely Morgan: Three-Wheelers 1910-1952
by Ken Hill
Completely Morgan: Four-Wheelers 1936-1968
by Ken Hill
Completely Morgan: Four-Wheelers from 1968
by Ken Hill
Daimler SP250 'Dart'
by Brian Long
Ferrari 250GT "Tour de France"
by John Starkey
Fiat & Abarth 124 Spider & Coupe
by John Tipler
Fiat & Abarth 500 & 600
by Malcolm Bobbitt
Lambretta Colour Family Album
by Andrea & David Sparrow
Lola T70
by John Starkey
Making MGs
by John Price Williams
Mazda MX5/Miata Enthusiast's Workshop Manual
by Rod Grainger & Pete Shoemark
Mini Cooper - The Real Thing!
by John Tipler
Morris Minor, The Secret Life of the
by Karen Pender
Motorcycling in the '50s
by Jeff Clew
Nuvolari: When Nuvolari Raced ...
by Valerio Moretti
Porsche 356
by Brian Long
Porsche 911 R, RS & RSR
by John Starkey
Rolls-Royce Silver Shadow, Bentley T-Series, Camargue & Corniche
by Malcolm Bobbitt
Singer Story: Cars, Commercial Vehicles, Bicycles & Motorcycles
by Kevin Atkinson
Triumph Motorcycles & the Meriden Factory
by Hughie Hancox
Triumph TR6
by William Kimberley
Vespa Colour Family Album
by Andrea & David Sparrow
Volkswagen Karmann Ghia
by Malcolm Bobbitt
VW Beetle Colour Family Album
by Andrea & David Sparrow
VW Beetle - The Rise from the Ashes of War
by Simin Parkinson
VW Bus Colour Family Album
by Andrea & David Sparrow
VW Type 2 Transporter Bus, Camper & Vans
by Malcolm Bobbitt

First published in 1997 by Veloce Publishing Plc., 33, Trinity Street, Dorchester, Dorset DT1 1TT, England. Fax: 01305 268864.

ISBN 1 874105 66 9

Readers with ideas for automotive books, or books on other transport or related hobby subjects, are invited to write to Veloce Publishing at the above address.

British Library Cataloguing in Publication Data -
A catalogue record for this book is available from the British Library.

Typesetting (Bookman), design and page make-up all by Veloce on AppleMac.

Printed & bound in England.

THE MGA

"First of a new line"

JOHN PRICE WILLIAMS

VELOCE PUBLISHING PLC
PUBLISHERS OF FINE AUTOMOTIVE BOOKS

MGA

"First of a new line"

ACKNOWLEDGEMENTS

Going back to trace events of several decades ago when the MGA went into production seemed a daunting task, but it was made immeasurably easier by the help I was willingly given by many people.

Gerald Palmer, Jim O'Neill, Terry Mitchell, Dennis Williams and Bryan Moylan - who were at Abingdon or Cowley at the time and played a major role - all provided information which has not been published before.

Another group of people, members of the MG Car Club, were equally helpful in steering me in the right direction, and I would like to thank in particular Paddy Willmer, editor of *Safety Fast!*, for pictures and access to John Thornley's memos to the BMC management - "tickling the leopard's nose with a straw", as he used to call it. Thanks, too, to other MGCC members - Geoff Barron of the MGA Register for reading the proofs, Bob West for details of historic MGAs, Andy Sargent, Rob Higgins and John Eagle, and to Nick Cox for advice on Twin Cams.

I have always found it strange that some books on British sports cars ignore the fact that the vast majority were exported and still have an enthusiastic following abroad. Help with overseas logistics came from MGCC members Brian Woodhams on South Africa, Henry Stenning in Australia, Len Bonnay in Canada and Dick Knudsen in the USA.

At home, at the British Motor Industry Heritage Trust, Anders Clausager, Richard Brotherton and Karam Ram assisted with documents, statistics and pictures and, as always, the library staff of the National Motor Museum at Beaulieu - Annice Collett and Marie Tieche - found nothing too much trouble.

Other pictures and/or information came from John Digby, Mike Ellman-Brown, Peter Wood, Malcolm Green, Dave Culshaw, Pedr Davies, and Sylvie Mitchell at *Classic and Sportscar*.

I am particularly grateful to have been able to quote road test data from the publications mentioned in the bibliography. Past copies of magazines like *Motor, Autocar, Road & Track* and *Autosport* have conserved important information about the way we drove then.

To everyone who has helped, very many thanks.

John Price Williams
Eastbourne
England

MGA

"First of a new line"

CONTENTS

MGA

"First of a new line"

INTRODUCTION

It was a bitter disappointment. The sleek roadster, produced on the sly at Abingdon by MG, was proudly shown in 1952 for the first time to the management of the new British Motor Corporation - and was rejected.

The new head of the merged Nuffield and Austin groups, Len Lord, had no love for Nuffield products. Three days before the Abingdon men showed him their new MGA, a burgundy roadster registered HMO 6, he had done a deal to produce the Healey 100. Donald Healey's sensational car fitted the BMC strategy as it used the Austin A90 engine, a 2.6 litre unit fitted to the hard-to-sell Austin Atlantic. John Thornley, who became MG's general manager that year, recalled: "In Lord's eyes, the cars were so similar that the MGA could not go ahead". Similar in concept, perhaps, but the Healey would be in a different price sector, have a much larger engine, offer a lot more performance - and sell in smaller quantities.

In retrospect, Len Lord was wrong in 1952, as faltering sales of the TF were to prove. But, two years later, he changed his mind, a design office was opened again at Abingdon and out came the drawings of HMO 6, project EX 175, which went on to become one of the world's best-loved and most successful sports cars. This is its story.

MGA

"First of a new line"

1

THE BEGINNING

Len Lord - a bitter old tyrant who hated Morris, MG and all the works of Nuffield; so say those MGA devotees who blame him for the fact that one of the world's favourite sports cars was launched three years later than it should have been ...

Len Lord and William Morris had fallen out in 1936 when Morris asked Lord to run the Nuffield empire, but refused him a share of the business. Lord went to Austin and by the time the MGA story really begins in 1952, he was head of the newly-created BMC conglomerate - the world's largest vehicle manufacturer outside America, producing 300,000 units a year. Of this figure, MG sports car production was a tiny part - 10,838 TD Midgets in that year.

In an oft-quoted remark, Lord promised to take Nuffield's headquarters at Cowley apart "brick by bloody brick", and this animus is blamed for his favouring of the Austin-Healey over the MGA prototype. But there was more to it than that.

Lord wanted a sporty Austin car he could sell in North America in those "export or die" times. His bulbous A90 Atlantic, built for that very purpose between 1949 and 1952, was a failure. He then tried to enter the same market with a pretty Jensen-built convertible on the A40 chassis. Same result ...

There is a theory that Lord then staged a 'beauty contest', approaching three small manufacturers and asking them to produce a sports car based on A90 components. Healey produced the 100, Frazer-Nash built a prototype based on the Targa Florio and Jensen designed a sleek two-seater. However, it was more complicated than that: in 1951 Donald Healey was already designing a new, mid-range sports car to fit the price and performance gap between the Jaguar XK120 and the MG TD Midget. The Riley engines Healey had been using up until then were to be phased out, so he approached Lord and asked for a supply of the A90 engines - a deal (which was also to include Austin front suspensions and axles) done on November 27 1951, according to his son, Geoffrey.

The prototype was built at the Healey works at The Cape, Warwick, and the body hand-made by Tickford. The car was tested at 111mph on a Belgian motorway and appeared at the Motor Show at Earls Court in October 1952. Healey planned to price the car at £850 plus tax and make a maximum of 25 a week.

The deal with BMC was done at the show. Healey's pale blue prototype was snapped up by Len Lord, who pointed out that he could make 200 a week on a special line at Longbridge and sell it at £100 less than Healey's price. Jensen got the contract to make the body, a new badge - adding Austin to Healey - was produced literally overnight by a Birmingham manufacturer, and the BN1 was born. More than 10,000 examples would be built before the MGA arrived.

The MG prototype stood no chance when it was shown to Lord three days later. Jim O'Neill, MG's chief body designer, thinks that Abingdon should have shown Lord the car sooner, which might have influenced the final deci-

sion, though this seems unlikely, given Lord's insistence on Austin branding and the use of A90 components. Thornley was later to say, tactfully, that the cancellation was due to tooling costs, and point out that the HMO 6 body was a much more complicated structure than the T-type body, once described as "two sheets of tin and a bundle of firewood".

Could there have been two new sports cars at the same time? Unlikely, given the capital available and the marketing problems it would entail. "It would have been crazy to introduce the MGA as a competitor then" said O'Neill.

Though BMC hardly shied away from having a multiplicity of models, it is likely that a further reason Lord rejected the car was because MG was still successfully selling the TD which had been launched two years earlier. In fact, 1952 was a record year for MG production; 10,838 TDs was a huge increase on the previous year's figure, though to BMC bosses MG was still a cottage industry in a small riverside town in Berkshire (later county boundary changes meant Abingdon moved to Oxfordshire). The TD was also a much cheaper car than the Healey, at £530 before tax rather than £750.

Despite the continuing (and to some surprising) success of the latest incarnation of the thirties Midget, it was plain to those at Abingdon that a radically different successor was needed. As is often the case in motoring, this evolved from racing - improving the breed.

Though there was no competitions department until 1955, and therefore no official support for racing, MG had quietly helped private entrants like Dick Jacobs, Ted Lund and George Phillips to prepare and maintain their T-series cars for races in Britain and the continent. The significance to the MGA story of this support is that, in 1951, a special car was prepared for Phillips which defined the shape of the production car that was to come four years later.

Phillips, a despatch rider who became *Autosport's* chief photographer, had some success as a private entrant at Le Mans in 1949 and 1950, which encouraged MG to offer him more than the usual support in the 1951 race. Phillips was to drive a special car, a TD with a lightened chassis and aerodynamic body. Although designated EX 172 on Abingdon's experimental register, it became better-known by its registration number, UMG 400.

The inspiration behind this car - and many of MG's greatest achievements - was Sydney Enever, a self-taught, intuitive engineer who joined MG in 1920 at the age of 14 and rose to become the company's chief engineer. Enever was the genius behind MG's record-breaking sorties in the 1930s, including the design of Goldie Gardner's streamlined 200mph K3 Magnette, known to the factory as EX 135. It was the aerodynamic shape of EX 135, combined with more contemporary styling from cars like the Jaguar XK120 and the Alvis TB14 sports tourer, which provided the basis for the Phillips Le Mans car and the template for the MGA. These basic lines were eventually reproduced, with some changes, in 100,000 cars.

UMG 400 was engineered and designed by Enever. The lines were his and the body was rumoured to have been put together by a 'tin basher' at Abingdon, but it was more likely to

The car which began it all ... Sydney Enever's inspired design for the George Phillips racer had all the styling cues which can be recognised in the MGA.

have been done on the quiet at the Morris Bodies branch in Coventry. The car was then assembled at Abingdon by Alec Hounslow, head of the development department.

Careful attention was paid to the aerodynamic properties of the body, constructed of aluminum panelling over wooden struts. During the war, when Abingdon stopped producing cars and was churning out tanks and aeroplane components, Enever worked with technicians from plane-makers Armstrong-Whitworth, so when UMG 400 was being created, he was able to use his contacts to secure time in Armstrong-Whitworth's wind tunnel at Coventry to test quarter-scale wooden models of the car.

Careful attention to air-flow paid off at Le Mans; on the Mulsanne Straight, the unsupercharged 1250cc XPAG engine from the TD took the car up to 116mph. But three hours into the race the top came off a piston, bent the valves and put Phillips and his co-driver, Alan Rippon, out of the race. Phillips was angry as he reckoned that a factory-prepared racing engine should be unburstable. MG were upset because they thought he had pushed the car too hard. They subsequently offered him the car, which Phillips refused with some dudgeon. Like many cars now considered historic but little-regarded then, it was broken up, "put under the hammer", as Phillips said. But it had achieved its purpose as a design objective and, to a certain extent, as a public relations exercise, for there was much mention of UMG in the motoring journals. Racing participation to whet the public's appetite for a new MG was to be used with even greater effect in the 1955 Le Mans.

In the meantime, after the qualified success of UMG 400 in the 1951 race, work quickly began on an MGA prototype based on its design. A blueprint of the body was ready by February 1952, with changes to the body sides to give a smoother line; the line of the front wing was raised as it carried past the doors, and the rear wings were larger.

There were further wind tunnel tests and it was probably as a result of

The rear wings were to be made larger and the line of the front wing raised as it swept into the door.

these that the decision was taken to site the characteristic engine bay ventilation grilles on the top of the front shroud. In early designs they had been on the sides of the wings, in Italianate style.

Several modifications to the slightly ungainly posture of UMG 400, in which the driver sat on top of the TD chassis, were intended. To reduce height, a new chassis was drawn by Sydney Enever, either on his kitchen table or in the back room of his home in Westminster Way, Oxford, depending on which MGA legend you believe. Sweeping the chassis rails outside the cockpit, which allowed the floor to be lowered between them, meant the driver's seating position could be dropped noticeably. George Phillips was later to claim credit for this innovation in a conversation with historian Doug Nye. Speaking of UMG 400 he said: "I didn't like sitting on the thing rather than sitting in it. I said 'Why don't you use a backbone chassis like the old R-type and sit the bloke on the floor'". (The R-type single-seater racer of 1935 had a chassis shaped like a tuning fork, with the engine mounted in the fork).

Sydney Enever's basic chassis design was given to Roy Brocklehurst, who was barely 20 years old, to refine. It was designed to take the TD power train because B-series components had yet to be put into production. Two chassis were built; one for the prototype and one as a spare.

The body was built at BMC's bodies branch at Quinton Road, Coventry, under the supervision of Eric Carter, who was to play a major role in the development of the MGA.

Three months later the car was finished and sprayed in deep red. It ran on trade plates until registered HMO 6 in November 1952, the month following its rejection. It became known at Abingdon (in those politically-incorrect days) as 'homosex', though was referred to in official circles as EX 175.

Both HMO 6 and its predecessor, UMG 400, had been developed virtually on the sly, and there are many reports that when S V Smith, the BMC board director responsible for Abingdon, spent his one day a week

The car BMC turned down. At a quick glance, it's an MGA, but the bonnet bulge necessary to cover the XPAG engine gives it away.

there on Tuesdays, the experimental cars were pushed into fields out of sight.

There was little to distinguish HMO 6 externally from the car which was to become the MGA. The windscreen was slightly shallower and there was a hump in the bonnet to clear the rocker cover of the XPAG engine. It is remarkable that Sydney Enever's design was right first time; little change was needed or made throughout the life of the MGA.

The Abingdon men were gravely disappointed at BMC's rejection of their prototype car. It is something of an irony that, in November 1957, production of the Healey was switched from Longbridge to Abingdon, where it was produced alongside the car it had delayed for so long. In the meantime, HMO 6 was pushed aside and the business of turning out T-types continued.

The TF stopgap

Three months later, in January 1953, the MG factory at Abingdon was approaching a major decision. In the dominant North American market orders for the TD Midget were still holding up well, but, given the antiquity of the basic design, which stemmed from the thirties, this situation could not be expected to last. New and faster competitors were threatening.

In the past, many accounts have said that orders were already falling fast at this stage, but MG records do not bear this out. In fact, an expected seasonal fall-off in the winter of 1952/53 did not materialise. Out of a total TD production of nearly 11,000 in 1952, the year of highest output, almost 10,000 cars had gone to North America. Just how important the North American market was at this time is illustrated by the total export figures - between 1949 and 1953 almost 30,000 TDs were built, of which more than 21,000 crossed the Atlantic. At one time, Abingdon was turning out 250 TDs a week, 200 for the USA and 50 for the rest of the world. The factory was well aware that its survival depended on American sales ...

John Thornley wrote to the BMC hierarchy: "Future production is inextricably bound up with the USA, and we must direct ourselves to this market first, last and all the time. If what we produce happens to be acceptable in the rest of the world, all well and good".

When the TD was introduced in 1949, its performance compared to the average American car was such that it could beat them at the lights and often on the open road. This, and its handling, endeared the car to the American driver. But since 1949, there had been a revolution in American power unit design which put the TD at a disadvantage. There was, remarked John Thornley, a "very marked clamour" from the US for a larger engine in the TD, and while the US market could probably be held during the summer of 1953, unless there was some change it was doubtful the rate of orders could be maintained.

Before the A ...

Apart from Gerald Palmer's car, there had been several design attempts by the Cowley staff to design successors to the Y-type and Midget.

Foreshadowing his proposal for alternative body styles on the same shell was an eighth-scale model of a very small car, one version of which, with telescopic front suspension, looked like a trials car. Two very ugly saloons - a four-door version of the Minor and a tarted-up Morris Oxford MO - went no further than the mock-up stages.

The first 2-seater sports car proposal was a modified Minor, probably with an engine which did not go into production, an 1100cc overhead cam version of the Wolseley 4/50 unit. Next, around 1948, were three versions of a 1.5-litre car, probably powered with the Riley engine - a saloon, coupé and 2-seater. The design was thought to be that of a Morris stylist named Blakeley. One of these cars was built, possibly the coupé, which ran around Abingdon finished in green. Because of a shortage of money the sills were made of wood.

At other times experiments were done with T-series cars using Austin and Riley engines.

A model of an early attempt to replace the T-series Midget. Note the alternative body styles that could be bolted on to a common shell. This, and the pictures below, were taken in the experimental shop at Morris Motors by Dennis Williams, who worked in the MG/Riley design office.

The proposed telescopic suspension can be clearly seen; it was not adopted for the MGA.

DO965 was a proposal in the late forties for a 1.5-litre coupé. An open two-seater version was also proposed which would have had a smaller wheelbase.

Tests had shown that the square shape of the TD meant that from 80mph more power had very little effect. It took some 50bhp to reach this speed, but it would need double the power to reach 100mph.

"The traditional MG Midget has had its day because of its very inefficient aerodynamic form ... we must now adopt the principle of 'streamlining'", said Thornley. But Abingdon had been refused permission to build the car.

The answer was to 'improve' the shape of the TD, without altering its essential appearance, and to use a larger version of the XPAG engine, a 1250cc unit which, in the TD, produced 54bhp at 5200rpm.

The endemic shortage of money at Abingdon meant that the new car could not be dissimilar to the TD - having the same XPAG engine, albeit slightly more powerful with 57bhp at 5500rpm, same running gear and same basic body structure, apart from the front.

The design work on the body was done in late 1952 and an attempt was made to give the nose a more streamlined look with a sloping radiator, and fake filler cap and faired-in headlamps in the front wings.

Jim O'Neill, chief body designer of the MGA said: "They knife-and-forked the TD into the TF at Abingdon. It was going completely the wrong way, angular style like that was finished". Nevertheless, on January 8 1953, a blue prototype of the replacement car - the TF - was shown to the BMC board, which approved it. (In the nineties Rover has skipped lightly over the lettering gap between the MGC and the new MGF and say MGD was a prototype replacement for the MGB and the MGE was the EX-E concept car shown at the Frankfurt show in 1985. To show how difficult these matters are, what will the MGF's replacement be called? How many horses will it have?) The logical next step in the alphabet, TE, was not taken lest the car became known as the tee-hee. It had enough problems without being regarded as an object of fun, though this did not bother Alvis who used TE for the 1963 Three-litre.

The larger 1500cc engine was not to materialise for another year, even though versions of the XPAG engine bored out to 1466cc had been tested before the TF was launched. The delay was said to be caused by problems of hitting water passages during siamesing of the cylinders in two pairs, and the expected arrival of the B-series engine of 1489cc, which was more powerful, scheduled for September 1953.

This engine, which was eventually to power the Magnette and the MGA, caused some perturbation at Abingdon. If there were any doubts about its reliability at introduction, it would not go in the TF, it was said, since conditions to which an MG engine were subjected - particularly in America - were far more severe than those in a passenger car. John Thornley didn't like it for several reasons (see the chapter on the B-series engine) and it was never used in the TF even though it was available. *Motor* confided that this was: "because it is still in fairly small production and it is not yet possible to build it in sufficient quantity for the large production of Midgets". While this may be so, Abingdon no doubt preferred staying with the devil they knew, the venerable XPAG engine of pre-war Morris 10 provenance, bored out to its limits by getting rid of water spaces and siamesing the bores.

It was, however, fanciful to suggest, as Thornley did to the board, that the performance of this engine in the TF would be "markedly superior" to the TD. He was later to write: "At 85mph or so, the TF finds the atmosphere almost as impenetrable as a brick wall".

The car was never released to the motoring press in Britain for road testing, allegedly because of its low performance compared with contemporaries, but tests in the US by *Road & Track* showed that the TF's maximum speed of 85mph was only four miles an hour faster than that of the TD, and the 0-60 figure only 0.2 seconds faster at 16.3 seconds.

There was another complication; while the cruelly-rejected MGA prototype, HMO 6, hung about in the wings, BMC was going ahead at Cowley with development of a proper Midget replacement. Nuffield's chief designer, Gerald Palmer, who was responsible for Riley and MG, had come up with an exceptionally attractive and unusual sports car. Some sources have described this as the Magna, a name last used in 1934 for the L-type six-cylinder, but Palmer does not remember it as having that name.

The new car offered two different wing and door types, which could be bolted to the common middle section. There was a separate wing type in the traditional sports car manner and a fully-integrated version, bearing a resemblance to MGB styling.

Contrary to what has been written elsewhere, Palmer has confirmed that it was an all-steel monocoque design, not a new body on the TF chassis, and it was not just a wooden styling exercise, as one car was built, powered by the B-series engine. Dennis Williams, who worked with Palmer in the Cowley design office, also remembered it as being of unitary construction.

"The American market wanted the traditional MG styling" Palmer said "and the Continentals wanted an all-enveloping body. I took the designs to Morris Bodies at Coventry, and I used to go up there and look at them putting the prototype body together. It was the separate wing model, which we then built up at Cowley".

So, at the start of 1953, Abingdon had the problem of trying to get a face-lifted TD into production as the TF, waiting for the new design to emerge

Palmer's 'traditional' MG roadster for the US market on the turntable at Cowley.

from Cowley, while secretly hoping that HMO 6 would become the new model.

The Palmer car was still at the drawing board and rough mock-up stage but was due to be ready for production by March 1954. This was alarming news for Enever and Thornley, who were convinced that HMO 6 was the car they wanted.

"Undoubtedly these motor-cars, or something like them, provide a very suitable long-term development from the TD", wrote Thornley, "but bearing in mind what was said about the need for sending to the USA only well-tried designs, we must be extremely careful not to place ourselves in the position where we are utterly dependent on the continuance of the market upon the premature introduction of an untried motor-car".

Abingdon doubted whether the new car could be ready by March 1954 and, if it were, would it be reliable and satisfactory, given the unorthodox design? The sensible course, therefore, they thought, was to introduce the face-lifted TD in September 1954 and let it run for 12 or 18 months: "during which time Mr Palmer's new motor-cars can be fully-developed and tried and all the bugs taken out of them".

This unenthusiastic welcome for the Cowley car was a strategy to play for time, while pushing the cause of what was described as "a fully-streamlined car, tentatively called the Series UA". This was, of course, nothing more than a euphemism for HMO 6, the rejected MGA prototype. It was intended to fill a gap in the US market between the TD and the Jaguar XK 120, and would be sold as a competition car and to those who wanted a really fast British car for long-distance travel. Produced in limited quantities, said the Abingdon submission soothingly, its price need be of no great concern and if introduced at the time of the TF, it would not lead the market to believe that it represented the ultimate shape of the quantity-produced MGs of the future. Years later John Thornley admitted that this was an attempt to get the MGA in by the back door.

Events elsewhere favoured his

The all-enveloping body was for continental markets. The windscreen had a clever swivelling mechanism which enabled the bottom of the screen to slide downwards and forwards over the bonnet rather than folding down flat.

Naming the MGA

What began as Series UA, became DO 1062, then EX 182 and finally MGA. UA was John Thornley's name for the project. DO was the prefix given to projects from the MG/Riley design office at Cowley, and EX the name for Abingdon experiments.

Before the war, MG had run through most of the alphabet from the C-type Midget in 1931 to the WA of 1939. In 1947 came the Y-type, followed by the ZA Magnette of 1953. The letter U had not been used, other than as a project name, and neither had X. The 'Magna' name, dating back to 1931 and which might have been the name of Gerald Palmer's 1953 car, was, like the car, abandoned, perhaps because it was too similar to the forthcoming Magnette.

The new car could, of course, have been called a TG, though this would have associated it with the unsuccessful TF. It was time to start again.

MG's general manager, John Thornley, had decided views about what cars should be called and was highly critical of BMC's advertising policy. The firm, he said, was in business to make and sell Austin, Morris, Wolseley, Riley and MG cars, but large amounts of money were spent advertising Westminsters, Oxfords, Pathfinders and the like - "names which are here today and quite likely gone tomorrow. As soon as one starts calling a thing a Cambridge, one stops calling it an Austin. If one day the Cambridge is superseded by a Wiltshire or a Merioneth all the money spent on Cambridge has gone down the drain. If the man in the street can describe any of our vehicles without using ... MG ... then our nomenclature is wrong".

Thus, as a result of this thinking, the MGA was born and, as Thornley said, it compelled the use of the name MG whenever it was mentioned. And a further 25 models could be produced before the same problem recurred!

Even then, it didn't all go Thornley's way. In a memo to BMC management in 1957 he was particularly critical of the company's American distributor, Hambro, who advertised "MG, Austin-Healey, Magnette". By this, he said, it was to be assumed that MG and Magnette were separate marques, so the success of the MGA did not, in the public's mind, rub off on the Magnette.

The other issue which continued to exercise him was the wasteful duplication of models in BMC and the policy of retaining the pre-merger sales networks, Austin and Nuffield, which competed against each other for sales in the UK, even in small towns of under 10,000 population. Though the duplication applied to the saloon cars rather than sports cars, the MGA, for instance, was stocked by the Morris dealer and the car made on the neighbouring production line at Abingdon, the Austin-Healey, sold by the Austin franchisee. It took another 30 years and the collapse of British Leyland to end what was obviously marketing nonsense.

The styling is less happy from the rear three-quarter view, though the treatment of the rear is neat.

Octagonal instrument dials - a styling cue from the thirties - were revived in the TF.

temporising tactics, for the Palmer car was abandoned. Palmer says it was because Abingdon was not keen on the car and wanted to produce its own design, which is perfectly true.

Jim O'Neill, who worked on the car at Cowley with Palmer, said many people were puzzled by the decision not to produce it. "When I moved to Abingdon I had a word with Syd Enever and asked him why; he just shrugged his shoulders and would not speak to me about it". One of the factors which played a part was Abingdon's "not invented here" syndrome, bolstered by the unhappy experience of having to build the Morris-inspired Y-type tourer, designed at Cowley, which put an ungainly tourer body on the YA chassis and pushed it along with the TC engine.

The Palmer car was also too far ahead of its time. Unitary construction, for what was then small-scale MG production, was just not feasible. Palmer remembered that his car ran around for a while -"I drove it, and it needed a lot of development" - then it was cut up.

The TF went into production in September 1953 and, a year later, the TF 1500 was rolled out - sold mainly in North America to keep the MG flag flying (one American cynic greeted the news of the increase in engine size with the remark: "So what? I could drink 216cc in one gulp". Apparently, he failed ...).

In June 1954, Len Lord finally authorised the MGA. The decision was forced by the cruel comparisons made between the venerable TF Midget and other British sports cars like the 115mph Austin-Healey 100/4 and the Triumph TR2. The Triumph had a basic price of only £5 more than the TF and gave 90bhp and more than 100mph.

MG's market in the United States was disappearing, halving in the years between 1952 and 1954. Vital export dollars were being lost.

S V Smith, the BMC board member responsible for Abingdon, told the factory it had finally got the car it wanted, a design office to put it together and then to cap it, a competitions department. The faith of the two guiding spirits, John Thornley and Sydney Enever, was justly rewarded.

MGA

"First of a new line"

2

THE ENGINEERING

The body

The BMC board changed from glacial slowness to indecent haste. "Suddenly it was all panic and scurry" recalled Sydney Enever. It was decreed that the car would be in production within a year - by April 1955 - an extremely short timetable.

The first drawings were done on June 6 1954, and were of modifications to the TF suspension and steering, which was to be fitted almost unchanged to the MGA, still described in the drawings book as a 'Midget'. Next to be devised were the sidemembers and fuel tank mounting.

Jim O'Neill, who had worked with Dick Burzi on the Austin Atlantic at Longbridge and then moved to Cowley, volunteered to join the new design office at Abingdon and had the job of productionising the body based on HMO 6. "HMO 6 was the MGA as near as makes no odds" he recalled, "we used it as the design. I did a quarter-scale body and took it to Morris Bodies in Coventry and from this we produced the full-size layout. We borrowed Don Butler from Cowley who did all the detailing."

"I reported every week to John Thornley at Abingdon telling him how things were going, but no-one from Cowley or Longbridge interfered with us and we made very good progress".

O'Neill stayed at Coventry for four months, putting off a hernia operation until the major work was complete.

The mock-up was done in wood rather than clay; the wooden pieces which comprised it were then used as bucks to hand-beat panels - such as doors - for the early prototypes.

Apart from the obvious body changes compared with HMO 6, like the omission of the bonnet hump for the now-redundant XPAG engine, there were other, subtler changes. The boot lid was bigger and the front wings slightly slimmer.

The car was finished in nine months - really fast work according to O'Neill - and the BMC deadline could

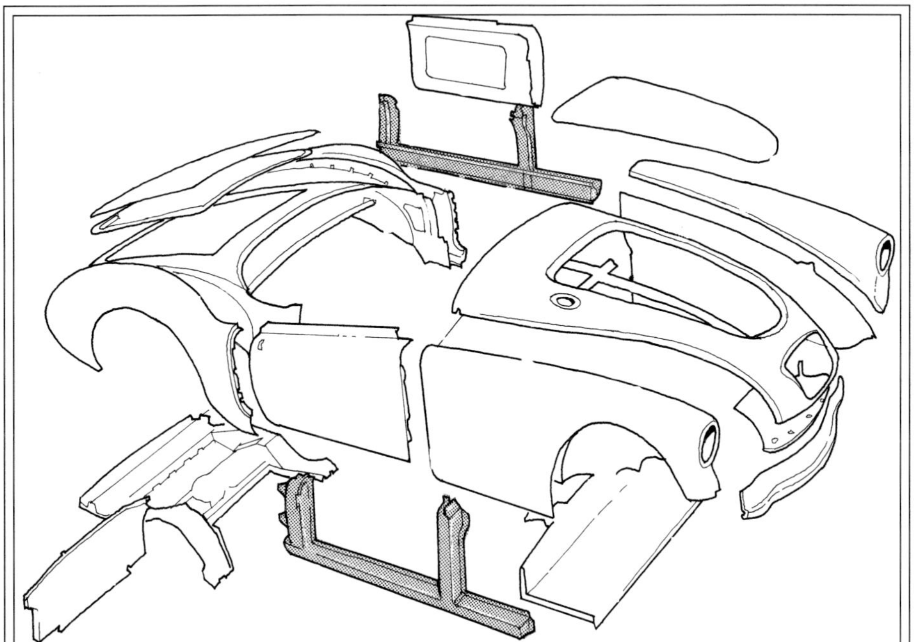

The main body panels were made by Pressed Steel and welded together at Morris Bodies. The highlighted structural members are the notoriously rust-prone 'F' sections on which the doors hang.

*Engines can be identified by the cubic capacity cast in the block - a quick reference for originality. As the MGA used an electric fuel pump, a metal plate (to the right of the number) was used to cover the aperture for the mechanical pump, which was still used on the Austin Cambridge. The 'MOWOG' stamped on the gearbox was a Nuffield designation for **MO**rris/**WO**lseley/M**G** components.*

have been met as the new B-series power train was available to replace the old Nuffield unit. But then came a snag - the production tooling for the body panels was not ready.

Pressed Steel had tendered for the job and suggested using new plastic tooling with which they were experimenting. They offered to provide the tooling free or, if it did not work, to build conventional tooling for an agreed price. "It just didn't stand up to the job", said O'Neill. "With a big pressing like the bonnet surround, they would get three or four off the press and then the tooling would break up. Maybe these days it would be different if they used carbon fibre". The unsuccessful experiment meant a reversion to cast iron tooling, which delayed the project by some months.

The first prototype body was delivered to Abingdon by lorry from Coventry, given its mechanicals and then tested by the head of the development department, Alec Hounslow. He is said to have remarked presciently on his return: "We'll sell a hundred thousand of these".

In the meantime, HMO 6 was utilised extensively for pre-production testing, not only for the new car but also, at a later date, for disc brake experiments, possibly those at the beginning of 1956 when Girling discs were tried on the *rear* wheels.

HMO 6 was eventually broken up on orders from Cowley - an event remembered with some sadness - and appreciation of motoring history's loss - by many Abingdon men .

The B-series engine

The engine intended to power the MGA would become one of the most successful ever built by a British company. The so-called B-series was used in hundreds of thousands of British cars from 1954 to 1980 and soldiered on until much later in commercial vehicles.

It was exported to other manufacturers; a typical example was a 1957 order from West Germany for 5000 to fit in Matador and Viking commercials. It lived on into the nineties in the Indian-built Morris Oxford, the Hindustan, and diesel and marine versions were available.

Work on producing a 1.5-litre unit had begun in 1951, the year that the Nuffield empire and Austin joined forces to become the British Motor Corporation. It soon became obvious that the Austin side of the Longbridge partnership, under Len Lord, was dominant.

Lord's plan was to replace the eleven different car engines used in the group with a more streamlined structure, a family of new overhead valve engines: the A-series to power the small cars (still used in the '90s in the Mini); the B-series the medium class, and so on to the largest capacity, the 4-litre D-series.

At that time, MG was still producing basically pre-war body designs - the TD roadster and the Y-type saloon - both using a pre-war engine, the famous XPAG engine of 1250cc, developed in 1939 to power the TB roadster, in which it gave 54bhp at 5200rpm. As described earlier, the prototype MGA which Len Lord rejected in 1952 would also have been powered by this engine, which was built at the Morris Engines factory in Coventry.

The B-series was undoubtedly an Austin engine, designed and built at Longbridge though its ancestry is clouded - some claim it descended from the Austin 10 of 1932, and although that was a 1125cc side-valve engine, it possessed the characteristics of later Austin and BMC practice in putting camshaft, manifolds and carburettor on one side of the engine and electrics on the other. It also had a stroke of 88.9mm, (3.5ins) which was to become a trademark of the B-series unit.

Some also see in the valve layout of the B-series the influence of Austin's first ohv unit, a pre-war truck engine, said to be based on a Chevrolet design. During the war, in order to build replacement engines for US Jeeps, it was planned to take this truck engine and reduce it to four cylinders. Though it never made much progress the engine was developed after the war to power the Austin Sixteen - the company's first ohv car engine.

The Ten engine, which produced only 32bhp at 4000rpm, begat in 1947 the 1200cc unit used with great success in the Austin A40 Devon and Dorset. This put out 40bhp at 4300rpm. It was designed by William Appleby (who later had a walk-on role in the twin-cam saga), but what was significant about the cylinder head was the work done for Austin in 1947 by Harry Weslake, an expert in gas flow technology and a close associate of Len Lord. Weslake redesigned the combustion chambers in the cylinder head of the Austin 16 to give greater efficiency, altering them to the heart shape which was to become another defining feature of not only the B-series, but also the A- and C-series engines. Weslake's experiments showed that this heart shape allowed a high degree of 'swirl' in the petrol/air mixture, giving good conditions for part-load economy.

The major work on the B-series engine design was done at Longbridge by Eric Bareham who first devised its dimensions in December 1951, less than a month after the Austin-Morris merger. His notebook, seen by Lindsay Porter and described in his book on the B-series, calls it the "Austin-Morris Engine. 'B' Series". The notebook shows several bore and stroke possibilities, from 1200cc up to and including "73.025 x 88.9 x 4 = 1489.34cc", the dimensions of the production engine that was to be used in MGs. Confusion often arises between the A40 engine and the B-series as they look similar, but the B, as might be expected, is both longer and heavier, with a stronger crank and larger bearings.

Hubert Charles, who had been MG's chief designer in the thirties, once described the B-series as a "horrible engine"; Eric Bareham called it "theoretically horrible" and John Thornley didn't like it much either! The reason for their dislike was probably the fact that the pushrods and ports were on the same side of the engine, making it very crowded. The ports had to be 'siamesed' to allow room for the pushrods, so cylinders 1 and 2 shared one inlet port, and 3 and 4 another. Cylinders 1 and 4 had their own exhaust ports at either end of the block, whilst the middle two cylinders shared

The engine location allows easy access to most regularly serviced components. The identification number is stamped on a plate between cylinders two and three and the chassis number on the bulkhead plate to the left.

one exhaust tract in the centre. As Lindsay Porter points out, it was only theoretically horrible because, whilst in abstract it might have led to induction problems because of the firing order, it actually worked very well.

Both the block and cylinder head were cast iron; the crankshaft forged in EN16 steel ran in three main bearings made of lead indium.

Prototype engines designed to produce 50bhp in the range of family cars were extensively tested in France and Spain, before the B-series appeared for the first time in the MG Magnette in October 1953, its 1498cc producing 60bhp due to a special manifold and twin SU semi-downdraught carburettors.

Further development in the six months before the engine appeared in Austin and Morris saloons resulted in several changes - a redesigned cylinder head with stronger valve springs and an increase in the length of the rocker arm ball pin - changes made as a result of experience at Abingdon.

Plans for the engine to go into the new MGA were begun in early 1954. Pistons for an 8:15 compression ratio were specified, compared with 7.2:1 for the Magnette, though this would be increased to 8.3:1 when the car was finally built. The pistons were solid-skirted rather than the lighter split-skirted versions used in other BMC cars and they had thinner piston rings. Larger valves with double springs were used to aid better breathing.

By the time the MGA was in production in the summer of 1955, the lines at Longbridge were producing more than 4500 B-series engines every week which were fitted to six of BMC's eleven models. The units destined for Abingdon were finished in dark red or maroon. There was one major change compared with engines for the rest of the BMC empire: the provision of a rev counter drive from the back of the camshaft. Power was increased quite quickly from the original 68bhp to 72bhp and the compression ratio from 8.15 to 8.3:1

Even as early as 1956 there were plans to open the engine out to 1700cc, not something which appealed to John Thornley. In a confidential memorandum to the BMC chairman, George Harriman, in April 1956, he argued for BMC's cars to be made lighter, rather than being powered by bigger engines. "We are not in immediate need of increased performance ... weight is the enemy. The new B-series engine will still be needed, but as future cars (and engines!) *must* be made lighter, it need have no larger a swept volume than at present".

He was, of course, talking about successors to the MGA, regarded from the outset as a heavy car for its engine capacity. It's ironic that the MGA, with its substantial chassis, was lighter than the car which followed it, the MGB.

The B-series was an engine which served MG well; the 1489cc original went into the MGA 1500, and the block was bored out to 1588cc for the bottom end of the Twin Cam. A pushrod version of this capacity was fitted to the MGA 1600 in 1959 - it was the only instance of this size of engine being used in the entire BMC empire.

Then, in April 1961, came the MGA 1600 MarkII with the 1622cc unit, the first BMC car to get it.

When the MGB went into production in 1962, the engine was bored out yet again to its final dimension of 1798cc. It has been said that this was because the heavier MGB was not fast enough with the MGA's 1622cc unit with which early prototypes were fitted. Although the MGB was the first of BMC's cars to be fitted with this 1798cc engine, it seems unlikely that the corporate engine policy in the medium car sector was dictated by Abingdon's requirements, and that the bored-out version was really destined for the Austin/Morris 1800 series.

The B-series engine was to gain another two main bearings and power the MGB to the end of its life in 1980 - not bad for a unit whose original design life was intended to be just five years.

MGA B-series engine data

	Type	Bore/stroke	Capacity	Output
MGA 1500 May 1955 - Jan 1959	15GB	73.025x88.9mm	1489cc	68/72bhp @ 5500rpm
MGA 1500 Jan 1959 - May 1959	15GD*	73.025x88.9mm	1489cc	72bhp @ 5500rpm
MGA Twin Cam Apr 1958 - May 1960	BC16GB	75.39x88.9mm	1588cc	108bhp @ 6700rpm
MGA 1600 May 1959- March 1961	16GA	75.39x88.9mm	1588cc	79.7bhp @ 5600rpm
MGA 1600 Mark II March 1961 - May 1962	16GC	76.2x88.9mm	1622cc	86bhp @ 5500rpm

**Changes to starter motor position and gearbox modifications. There was no alteration in bore size or power.*

It was fortunate that the 1953 design was over-specified and, despite the bores always being close together, there was enough metal to find another 309cc by widening the bores at various times, though this did mean that eventually water jackets between cylinders 1 and 2 and cylinders 3 and 4 were deleted.

Unlike the engines used in some sports cars, the B-series was never temperamental and, due to Austin's insistence on first-grade materials for components like cranks and valves, it not only had a long and reliable life but also proved surprisingly tuneable.

The decision to specify expensive lead-indium bearings in the MGA was vindicated when four BMC cars fitted with the B-series engine underwent extensive testing on the autobahn between Stuttgart and Heilbronn in the autumn of 1956. Doing between 600 and 800 miles a day at 60mph, the thin-wall white metal bearings collapsed within 9000 miles, whilst the lead-indium bearings were unscathed after 25,000 miles. The trials also led to the fitting of full-flow oil filters on other BMC cars.

Elegant it wasn't, but Len Lord's B-series long-stroke plodder, given the Abingdon treatment, provided more than half a million MG sports cars with their fun and fizz.

The chassis

In many ways, the MGA was technically out of date before it was built. By 1955, very few brand new models were still being built on a separate chassis with wooden floorboards (Nuffield had begun production of a unitary chassis/body as far back as 1939 with the Morris 10 Series M), although Abingdon was to produce the Riley Pathfinder on a separate chassis until 1957, Triumph adopted the same configuration for the Herald as late as 1959 and Morgan still do so. The reason was not one of failing to accept new methods, but usually a lack of money in smaller firms to build the press tools necessary for full monocoque construction. Alvis, for example, failed totally in trying to build a unitary construction car in 1949, and again in 1954 when a new Issigonis-designed car failed because they could not afford the tooling costs at Pressed Steel. This forced them to use the same chassis from 1950 until 1967, when production ceased.

The high cost of unitary construction in the early fifties became viable only with large production runs of saloons. The Morris 10 was Nuffield's first unitary car, followed by the Minor in 1948. In the low output sports car sector, it was cheaper to build the traditional way. Remember that in 1954, when the MGA was being planned, Abingdon made only 6500 MG TFs, whilst Minor production at Cowley was averaging ten times that number.

Though Gerald Palmer at Cowley had designed a monocoque replacement for the TD, it was rejected. The reason, he says, was that the people at Abingdon wanted to carry on with the traditional chassis frame, which was true, but there was also a political dimension. A unitary-construction body would probably have to have been made at Nuffield Metal Products at Cowley, who made the Minor. They had the presses to do it, and a shift of MG production to either them or Pressed Steel would have left Morris Bodies at Coventry very short of work, and even Nuffield Metal Products was not geared up to make what might have been small, expensive runs of MG monocoque shells.

However, the planners could scarcely have imagined that the MGA would go on to sell more than 100,000 examples, and so, by the late fifties, unitary technology was beginning to spread into sports car construction as more plant came on stream. The Austin-Healey Sprite in 1958, and the Sunbeam Alpine of the following year, used monocoque shells and very quickly the use of separate chassis frames came to an end. Just how

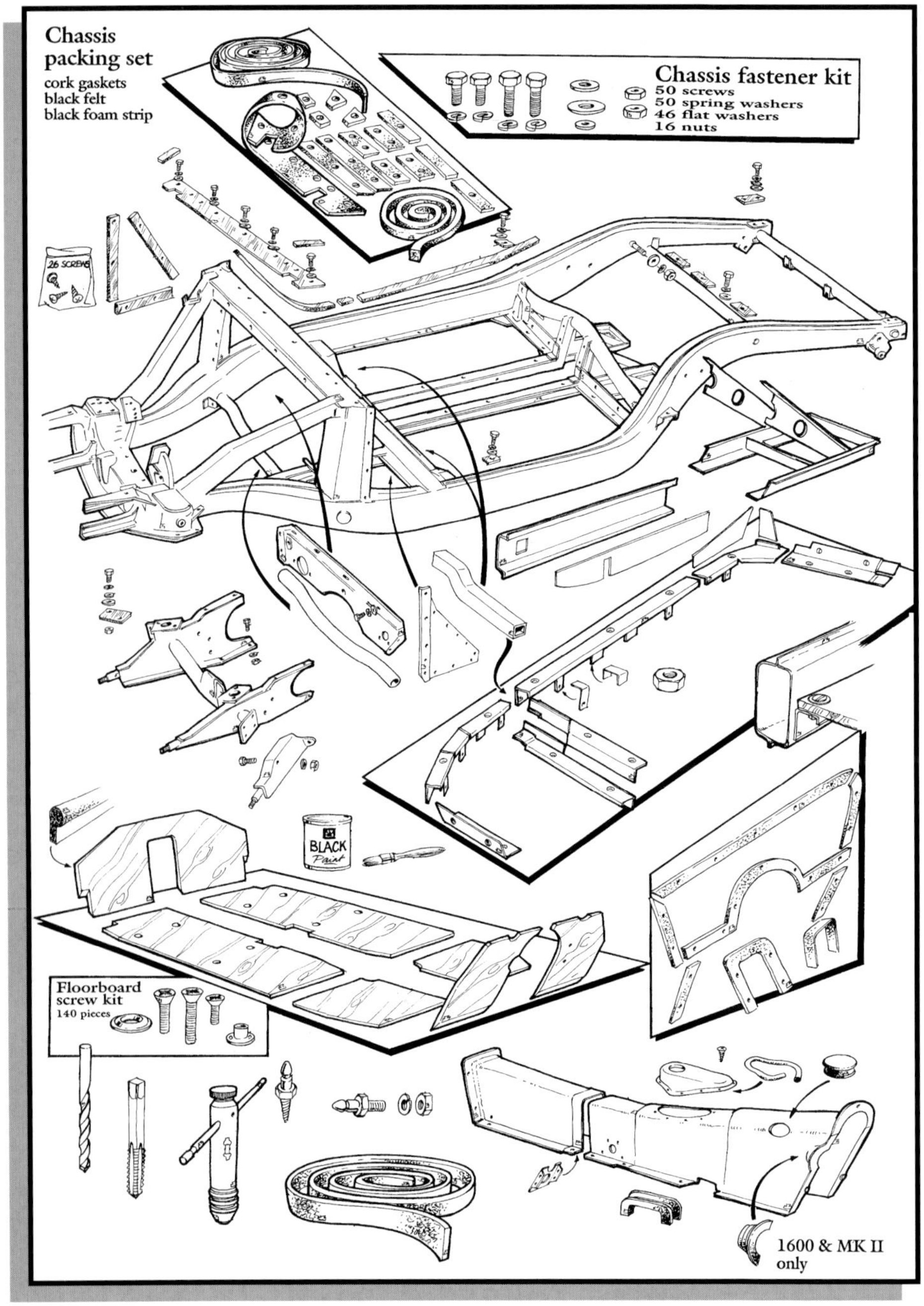

MGA chassis, floors and gearbox cover (Courtesy Anglo Parts, Belgium)

quickly is illustrated by a note which Sydney Evever and John Thornley wrote about MG's future plans after visiting the London Motor Show for the MGA's 1955 debut: "The great disadvantage of a monocoque form, particularly in the case of relatively small production rates such as our own ... is that it is necessary to tie oneself to a body design too far ahead of production. By using a self-supporting chassis, development of body and chassis can proceed independently.

" The complete design then enjoys the benefits of flexibility, such that the style may subsequently be changed without interfering with the chassis and vice versa.

"We therefore consider that all future Abingdon products should have chassis frames".

Thirty-six years later, in *Safety Fast!*, the MG Car Club Magazine, Thornley remarked wryly " Within a little more than three months [of writing the above] we were deep into the design of what was to become the MGB - which is, of course, monocoque! Heigh ho! You can't win them all!"

But certainly in 1954, when the MGA was being hurried into production, there seemed no alternative to a chassis. Talking to people who worked at MG in the fifties, it becomes apparent that money was always very tight and that Abingdon's needs were low on BMC's priority list. "MG was always the poor man of the outfit", said Jim O'Neill, "and it was so much easier then to have a cheap chassis frame than a monocoque vehicle".

John Thornley had considered not only the conventional box section chassis but also a Porsche-type space frame chassis which could accept alternative body skins. A month after the MGA was launched in Britain he wrote: "A purely experimental tubular chassis, with quarter elliptic rear suspension, is already in being. There is ample room within the MGA bodyshell for substantial changes in chassis layout. A more advanced shape is already being worked on against the day when this might be needed".

Terry Mitchell, chief chassis designer, recalled he built a tubular chassis for the MGA, but what happened to the car is a mystery. "We used to do lots of experiments like that which never

Driver and passenger sit inside the chassis rails rather than on top of them, as had been the case with previous MG sports cars. The practice of wiring together two 6-volt batteries continued into MGB production.

saw the light of day". This was one of the cars built for the 1956 Le Mans race that did not run because of BMC's withdrawal from racing. It was suggested the concept had been abandoned because of build quality, for *The Motor* noted in September 1955 that the new car's chassis was conventional and that: "the 'space frame' used to provide rigidity with less weight on some competition models having been rejected as incompatible with good doors". This meant, presumably, that there was not enough support in such a frame to maintain proper shut lines.

But it was economic factors which led to the MGA chassis being produced as MG sports cars always had been - on a robust metal chassis pushed by hand along the production line to be mated with its engine and body. The chassis was arc-welded together - the sidemembers and six crossmembers, welded at attachment points, were made largely of substantial 14-gauge steel. The two sidemembers were boxed throughout their length and spaced to the full width of the body, increased in section at stress points and sweeping inwards towards the front to allow for a good turning circle. At the rear, they swept over the rear axle, a design first used in the TD; before then MG chassis were underslung, or below the rear axle, which restricted rebound movement.

Behind the engine, the substantial 'goal post' structure, braced front and rear to the main members, ensured that scuttle shake - the bane of sports car designers - was reduced to a minimum. It also had the effect of making the front end very stiff.

Jim O'Neill recalled that this type of structure was a legacy of the TD, which he designed in 1949 at the age of 28. "When the TD prototype came back from Morris Bodies branch, I was really proud of it, but you couldn't read the instruments because the scuttle shake was so bad. I hadn't a clue what to do.

"No-one at Cowley [where the MG design office was then located] was much interested but the chief draughtsman, Tom Ramsay, sat down with me and we designed a 10 gauge vertical plate and welded it to the chassis, then built it up into a goalpost. The tubular structure cured the shake and it was carried over into the TF and the MGA".

Although the received belief is that the MGA chassis was a development of the Phillips Le Mans car, with occupants sitting within the full-width frame rather than on top of it as they had done in the Midgets, the idea had originally come from the United States, where the so-called 'periphery' frame had been commonly adopted. It was

The front suspension and the rack and pinion steering gave excellent roadholding.

The suspension was, indeed, 'specially developed', but for a Morris in 1938 - some 17 years before the MGA was launched.

so-named because the outside chassis members followed the body contours, or periphery, of the car. The principle had recently been established at Abingdon, where Gerald Palmer's new Riley Pathfinder had just such a chassis.

The basic MGA chassis shape had been used for the ill-starred HMO 6 in 1952. Terry Mitchell said that it needed little modification for the production version two years later. It was shortened by an inch, the front track made wider and the rear narrower.

He remembered that they began the chassis work in July 1954. He worked on the rear suspension, Harry White, then chief chassis designer, did the 'goalpost', which changed from a tubular to an angular structure, and the front suspension was done by Jack Daniels, a close associate of Alec Issigonis.

Both front and rear suspension were hardly new; in fact, the coil spring independent front suspension was designed by Issigonis in 1938 and was first used by Abingdon on the MG YA saloon in 1947 and then on the TD.

Another bequest of the YA and TD front ends was the excellent rack and pinion steering made by Cam Gears, which was far superior to the Bishop cam used on earlier sports cars.

The top wishbones were pivoted on Armstrong lever-arm hydraulic dampers; MG stuck to this configuration despite the move by the rest of the industry towards telescopic dampers. At the rear, the half-elliptic leaf springs were dampened by double-piston lever arms - again, pre-war practice - but this suspension set-up served MG so well that it persisted until the last MGB.

Pre-war tradition also dictated that the batteries, two 6-volt Lucas SG9E installed in series, were mounted side-by-side at the rear in cages welded to the chassis either side of the propeller shaft. This practice, previously abandoned in 1939 with the demise of the TB, had the advantage of lowering the centre of gravity, reducing weight on the front wheels and freeing space in the engine compartment, and persisted through MGA and into MGB production until September 1974, when two batteries were replaced by a single 12-volt battery)

Torsional testing of the chassis was done at Cowley by Alec Hounslow and Henry Stone, both of whom had been closely involved with the Le Mans cars.

The chassis had also been tested under extreme conditions. In 1954 preparations were being made for another record-breaking attempt in the United States, which is dealt with later

A heavily-stylised chassis at the London Motor Show, with sectioned engine displaying the interior.

in the book. The relevance of this to the development of the MGA is that the new chassis designed for HMO 6 was to be used in the attempt.

HMO 6 was tested with a cockpit cover, spats and undershield, but was not aerodynamically efficient enough. It was here that the second chassis, built at the same time as the original, came into play. A new streamlined body was built and the car, project EX 179, went on to take dozens of records.

All in all, the chassis was a substantial piece of work. Its weight was noticed by road testers from the car's introduction and, whilst it made the car slower than it might have been with a flimsier structure, it did give it first-class torsional rigidity, resulting in excellent roadholding. Sydney Enever once said that the chassis could rust away two thicknesses and the MGA would still be a good safe car to have a 'shunt' in. His successor, Roy Brocklehurst, did indeed have such a shunt in one, putting the third production car built, LJB 370, into a ditch in 1959 - without injury. He once mischievously described the frame to a motoring journalist as "probably the best separate chassis ever engineered" - omitting to mention that he had largely been responsible for creating it from Enever's design.

MGA

"First of a new line"

3

LE MANS - AND THE START OF PRODUCTION

EX 182

The new MGA's first public outing was to Le Mans on June 11 1955, when three cars in British Racing Green took to the Sarthe circuit. It was made very clear to the motoring press that, whilst these might be specially-prepared cars, they were, in fact, very close in detail to a production car about to be announced. John Thornley had decided a year before the race that substantially standard MGAs would compete, though he did not, of course, know then that the production car would not be ready.

His intention, at the 1955 Le Mans, was to give a demonstration of high-speed reliability of a new model to qualify for the Biennial Cup the following year. This could be done only once. "In 1956 we must enter to win our class, and a pretty stiff class it now is." he wrote in a memo to BMC management. He was adamant that this should be achieved with cars near to production versions. "There is no point in building five or six ultra-special cars for racing. This is very expensive, proves very little and the competitions for which they can be entered are very few".

Le Mans was not only a debut for the car, it also marked the official return to racing of MG works cars, which had stopped racing in 1935, supposedly as part of an agreement between Nuffield and Austin to cease competing against each other, although record attempts and unofficial race sponsoring were common both before and after the war.

Le Mans in 1935 had been the last works outing, when an all-woman team in P-type Midgets finished 24th, 25th and 26th. Now, 20 years later, the racing ban having being lifted in December 1954 by BMC's top management, MG was back on the Sarthe circuit. And not only that; BMC belatedly realised that competition was good for sales and made Abingdon the headquarters of the newly-formed BMC Competition Department, whose job it was to publicise the company's major racing, rallying and record-breaking products.

In charge of the workshops was Alec Hounslow, a former riding mechanic to Nuvolari who had spent a short but significant time before the war with the Squire company, whose perfectionist founder tried to build his British sports cars to Bugatti standards. Under Hounslow, the re-opened racing department made meticulous preparations for Le Mans and the cars received the sort of attention which no production car ever did. For instance, Hounslow insisted that every nut and bolt on the cars was lock-wired to prevent components working loose during the gruelling race.

The project in MG's experimental register was called EX 182. It took the production chassis and body shape of

Left: A hand-formed Le Mans body outside the works, showing the removable metal cover over the passenger seat and the fairing for the rear view mirror. (Courtesy BMIHT/ Rover Group)

Right: The racing chassis floor and cover for the propeller shaft was panelled in aluminium. (Courtesy BMIHT/Rover Group)

the MGA, but clothed it in 18-gauge aluminium alloy secured to the frame by countersunk rivets. The aluminium panels were produced by Morris Bodies on the MGA bucks, then finished and attached at Abingdon. They featured several removable panels, held by quick-release Dzus fasteners, such as the aerodynamic undertrays to reduce drag and a one-piece aluminium gearbox tunnel. As the plan was to use the cars subsequently in the Alpine rally, the passenger door, permanently closed at Le Mans, would be made openable and a sheet of alloy - which was quickly detachable - covered the passenger side of the cabin.

Mechanically, the units were standard B-series engine, gearbox and rear axle, as were then fitted at Abingdon to the MG Magnette ZA, though with modifications to allow unremitting high speed work. The connecting rod and big end bearings were lead bronze rather than white metal, the compression ratio was raised to 9.4:1 and there was a modified cast iron cylinder head. Much work had been done on this by Harry Weslake, the air flow consultant to BMC.

What distinguished the installation from the normal B-series was that the inlet ports were carried right across the engine and joined together on the ignition side by a balance pipe. The point of this was to try to maintain a steadier flow of mixture from the two SU 1.75 inch carburettors because with siamesed ports there are asymmetrical flow conditions in the manifold.

With this layout, the carburettors could, in theory, have been installed together on either side of the engine. There has also been a suggestion that there could have been a set of carburettors on *both* sides of the engine: one set for slow running and the other for

Setting off for France ... the tiny competition windscreens were small protection against wind and weather, so goggles were a necessity, as there was a violent storm on the road between Boulogne and Le Mans.

high speeds. However, any benefits from either layout seem to have been slight for the standard SU installation, on the right-hand side of the engine, was adopted and the balance pipe on the left was not used on the production car.

The carburettors were fed from a 20-gallon tank between the chassis rails at the rear by two SU high pressure fuel pumps, the filler cap protruding from the centre of the boot.

The modifications, with changes to cam profiles, gave an indicated 82.5bhp at 6000rpm, compared with the modest 60bhp from the Magnette installation and 68bhp in the MGA at its introduction.

Other mechanical changes included a close-ratio gearbox and a special high back axle ratio of 3.7 to 1 - needed to attain maximum speeds on the long straight. The car weighed 1596lb (725.4 kilos) compared with the first production MGA weight of 1904lbs (865 kilos).

Twelve of the Le Mans engines were built and, to avoid the possibility of breakdown through heat build-up, head gaskets were not used. The cylinder head and block were painstakingly lapped together by hand, carborundum paste between the surfaces, rotated for hour after hour by hand until a perfect metal-to-metal, gas-tight fit was achieved.

Special oil coolers were built in the racing department of Morris Radiators by Roy Towersey and, at one stage, there was even a proposal to cool the rear axle oil. Brake drums and backing plates were drilled for lightness and cooling.

The engines were run on the test bed for 24 hours before being taken apart again to check for damage.

By April 1955 the cars were ready and were tested at Silverstone. Marcus Chambers, MG's competition manager, recalls in his memoirs *Works Wonders* that the test drivers - who would later drive at Le Mans - made some criticism of the handling and driving positions, but nothing which could not easily be rectified.

One of the cars was sent to Le Mans to try out gearbox and final drive ratios.

Two cars were originally accepted for the race in the 1.5-litre class production category, but because of the delay in bringing out the road car, they were transferred into the prototype class (allowing them to use the alloy body rather than the much heavier steel version); a third car on the reserve list at number six was accepted just before the race. A fourth was to be taken along as a spare.

The line-up was:

Car No:	Engine:	Chassis	Drivers
41	182/42	182/38	Ken Miles/Johnny Lockett
42	182/43	182/39	Dick Jacobs/Joe Flynn
64	182/44	182/40	Ted Lund/Hans Waeffler
40	182/45	182/41	Spare car - unused

The cars were registered LBL 301 to LBL 304, but since the factory frequently swapped around numbers, it is not possible to ascribe these to any particular car, though it is widely assumed that they followed the numerical sequence of chassis 38 being LBL 301 and so on.

Three of the cars were driven to the south coast by Alec Hounslow, Doug Watts and Jimmy Cox, then ferried to Boulogne for the last stretch to Le Mans. Behind them came the mobile workshop/transporter - specially-built for the competition department by Appleyard of Leeds - which contained the spare car. The rest of the entourage, including a doctor, travelled in three Magnettes and a Riley Pathfinder.

George Eyston, breaker of many records in MGs and later president of the MG Car Club, had arranged for the entire team to stay at the Château Chêne de Coeur near the circuit, home of the Vicomtesse de Vatibault, where there was not only room for them, but stables which were used as workshops and a gravel drive where the drivers practised their Le Mans running starts.

In the camaraderie of the pre-race preparations, no-one could have guessed at the tragedy which was to come. "The black shadow of horror destroyed the 23rd 24 Hours Grand Prix of Endurance at Le Mans last weekend." reported *The Motor*. "In an appalling accident, the veteran French

The cars outside the chateau with the new BMC transporter. In the kitchen in the back of the transporter John Thornley's wife made meals for the team during the race.

driver Pierre Levegh lost his life at the wheel of a Mercedes 300SLR, more than 80 spectators died and over 100 were taken to hospital". Two hours into the race, Levegh's car plunged into the crowd after a collision with Lance Macklin's Austin-Healey. Dick Jacobs in MG car 42 saw smoke as he approached the grandstand. "I think this must have distracted me and I lost concentration", he wrote later. "I lost control, hit the embankment ... and got thrown out, skating along the road for some distance. I don't remember much after that".

The car was a write-off and Jacobs, a popular figure, was seriously injured - so much so the London papers thought he was dead and printed his obituary. The flag on the Abingdon factory was lowered to half-mast until news came through he had survived. Though he never raced again, he was much involved in MG motor sport in later years.

What seems inconceivable today is that, despite the carnage, the race was allowed to continue for its allotted length. The two remaining MGAs lapped steadily, with no problems apart from an accident to Lund's car when he ran into the back of a D-type Jaguar. The impact pushed the left-hand front wing on to the tyre and pulling it free required a six-minute pit stop.

It was John Thornley's intention that the cars should exceed 80mph on average, including pit stops, thus qualifying them for the Biennial Cup. In the event, Lockett and Miles in car 41 averaged 86.17mph over 248 laps finishing 12th overall and fifth in class. In car 64, Lund and Waeffler completed 230 laps at 81.97mph, finishing 17th out of the 21 cars which completed the race and sixth in class.

Though the tragedy and its aftermath - the banning of motor sport in France - cast a shadow over the event, the Abingdon team achieved a great deal in promoting the forthcoming MGA. The new streamlined shape of the cars, compared with the T-series, was widely praised and attention drawn to their reliability and top speed of 119.5mph, which car number 64 had achieved in the timed kilometre along the Mulsanne Straight.

The next major sporting event was the Alpine Rally a few weeks later, and plans were made to enter the cars again. But the rally was suddenly cancelled as a result of the Le Mans tragedy.

However, the factory decided that

The Lund-Waeffler car is pursued through the corners of the Sarthe circuit by number 41, which finished fifth in class, one place ahead.

Le Mans replica

Of the five original 1955 Le Mans cars, one survives - number 64 - the Ted Lund/Hans Waeffler car which finished 17th. It is in the Westminster MG car museum in Vermont, in the United States. The trim is not original since the cars were modified immediately after the race in preparation for the Alpine Rally. Such is the fascination of these cars, that one attempt has been made by an MG enthusiast to recreate the original specification and build a replica, with the help of many of the people - like Tommy Wellman, Cliff Bray, Jimmy Cox and Gerald Whiffen - who built the original race cars.

The painstaking project was completed in 1992 and took more than 2000 hours. A 1600 chassis was modified to bring it into line with 1500 specifications, then further altered to make it like that of LBL 301, with bracket modifications and only one battery tray. The engine was lightened and balanced, and the head and block lapped to enable the engine to run without a head gasket, following the practice used for the race cars. The SU long-throat carburettors were built by Don Law, who made the originals.

The aluminium body was hand-made; at one point the alteration of a flange by 0.25ins took 15 hours. Everything was copied as accurately as possible from the 1955 car, but some things could not be replicated: the alloy rims of the 62-spoke wheels were not available, neither was the Dunlop tyre tread pattern. The only other departures were cylinder head porting and the gearbox and back axle ratios which were made more suitable for road use.

The car was sold to a Yorkshire collector at an H&H auction at Buxton in Derbyshire in September 1995. It went for £13,900 ($21,000) - well below the cost of construction.

the outing would have been a good test bed for the new MGA and should continue in a different form. Taking a Le Mans car, a pre-production MGA prototype and a Magnette and Austin Cambridge fitted with Le Mans engines, a team of testers travelled more than 2000 miles in Europe, including visits to the tracks at Montlhéry and the Nürburgring. The tour resulted in changes to braking, suspension and ventilation.

Since the road car would not be launched until the Frankfurt Motor Show in September of 1955, the publicity offensive was kept up by allowing motoring magazines to road test the Le Mans cars in July.

Changes had already been made to the cars to prepare them for the Alpine Rally. More normal gear ratios were required, so the close-ratio box was replaced by the standard Magnette unit and the rear axle ratio lowered to 4.3:1 instead of 3.7:1. Alterations to the bodywork made the cars look even more like the production version to come. The small Le Mans perspex windscreen was replaced with a full-width screen with wipers, a sheet metal tonneau cover secured with Dzus fasteners over the passenger side to reduce wind drag was removed and hood and sidescreen were installed. It demonstrated how a racing car could be quickly transformed into a road-going sports car.

The original grille, with its aperture in the offside containing a Lucas Flamethrower light to pierce what is always described as the "notorious" early morning mist at Le Mans, was replaced by the grille which has become so familiar.

The press thrashed LBL 303 and loved it. The *Autocar* recorded a top speed of 103mph and a 0-60mph time of 13.8 secs. Its view was that if the engine could be offered without much detuning and the weight (now with all equipment up to 1817lbs 825.9 kilos) could be kept down "this delightful car will be a serious contender in the small high performance class and should have a ready sale in home and overseas markets".

Autosport did even better with LBL 304, probably the spare Le Mans car, recording a top speed of 104.6mph and a 0-60mph time of 12.8 secs and waxed even more lyrical, saying the conservative British public had finally 'had' the old-style sports car. And with a dig at the venerable T-series it went on: "At long last the public is willing to forget the sports-lorries of 1929 and this MG has just the sort of look that they will be after".

There was to be one more outing for the EX 182 cars, a week before the official MGA launch at the Frankfurt Motor Show in September 1955. A team of three - the two Le Mans finishers LBL 301 and LBL 303 and the spare car - were entered for the Ulster TT at Dundrod - events described later in the Twin Cam chapters. It was another widely-publicised display of the new MGA, associating the breed with speed and preparing the way for the production car.

Into production

Gaining permission to open a design office in Abingdon to produce the MGA was a significant step for MG; it meant that they were able to acquire some control of the sort of cars built- and not be dictated to by Nuffield headquarters at Cowley, which was always regarded as unsympathetic.

In the early fifties, about twelve people worked in the design office at Cowley, under Gerald Palmer's supervision, on MG cars and later on Rileys, after the death of Harry Rush. Several of those talented people, like Terry Mitchell and Jim O'Neill, now moved to Abingdon. O'Neill said: "I was very impressed with what John Thornley did; I'm sure he realised that unless design moved to Abingdon, MG would always be playing second fiddle to Cowley".

There was a lot of work to be done

No. 64, as modified after the race for road use. This was the car tested by Autocar *in July 1955, which described it as a serious contender in the small, high performance class if the weight could be kept down to its present level. In fact, LBL 303 was about 200lbs lighter than the production car.*

in a very short time. As Terry Mitchell, the chassis engineer, put it: "Drawing the outline of a prototype is the easy bit. It's what comes next that's difficult. Syd Enever was always drawing sketches on the back of cigarette packets and matchboxes turned inside out and bringing them in to you; he did tend to think you could read his brain - and after a while you could".

Putting the MGA production car on the road required some 1000 engineering drawings. It was a much more complex structure than the previous T-series with its little body tubs. Nevertheless, by mid-May 1955, despite delays at Pressed Steel, the first production cars, as opposed to Le Mans cars, were being produced.

It was notable that although some areas of the car suffered from penny-pinching, the decision was taken to skin the doors, bonnet and boot lid in aluminium to reduce weight, and this at a time when aluminium cost three times as much as steel!

There had been extensive testing of the pre-production MGAs, particularly at the nearby airfield at Halfpenny Green, where a skid pan was used. Suspension and tyre tests were done at the Dunlop track - modifications made as a result of these experiments were said to have added 3mph to the car's top speed.

Torsional testing of the chassis was done at Cowley, wind tunnel testing of the body at Longbridge and road tests were done at MIRA, the motor industry's research establishment. Here, the new car was put through its paces on the track and in the dust tunnel. It was also run there on *pavé*, the notorious rough cobbled surface still to be found at that time in France and, in particular, Belgium. The strength of the chassis proved its worth there, unlike those of the first Frogeye Austin-Healey Sprites, designed at the Healey works at Warwick. Terry Mitchell recalled the incident: "MG took over the Frogeye from Longbridge and Syd (Enever) said we'd better put a couple on the *pavé* at MIRA, so we took them up there and they both broke their backs."

The problem was the weight on the quarter-elliptic springs just behind the seats. "Syd stopped the production line and Harry White, the chief chassis designer before me, who'd done the front section of the MGA, was asked to come in at the weekend and quickly design some stiffeners. He wouldn't come in and left two weeks later to go to Rootes, which is how I got his job."

Dennis Williams, also there at the time, remembered that when the line was stopped there were about 100 Frogeyes in various states of completion. "We had to run the line backwards, stripping out everything we had put in so that the shells could be sent back to Pressed Steel to be modified. It wouldn't have happened if they had used half-elliptic springs as we did on the MGA."

The story is worth recounting, even in an MGA book, since it gave those at Abingdon a certain grim satisfaction and a feeling that nothing like it would ever have happened to a sports car designed there.

The thoroughness of the MGA's testing meant that during the seven years of production - Twin Cam apart - very few changes had to be made to the basic structure, though it seems that minor last-minute alterations continued up until launch.

Early studio publicity shots show the car with bumpers identical to those on HMO 6 and a bootlid without cross-bracing.

MG chassis numbers for each new model had traditionally begun at 251, the factory's telephone number, but the system was changed for the MGA, which followed directly on from the last TF 10100.

The first production MGA, chassis HDE 13 10101, was built between May 16 and 25 1955. It came off the chassis line on May 17, the trim line and body line on the 19th and passed out of the rectification department on May 25. It

The first six MGAs

10101 - registered LJB 205, right-hand drive car, finished in Tyrolite Green with grey upholstery. The chassis number was changed to EX 182/42 by the development department, following on immediately from the fourth Le Mans car, so this could well have been modified to the same specification.

10101 was then applied to LJB 439 the left-hand drive prototype, to North American specification, built in the competitions department. It was Orient Red with black trim and given the same chassis number, but a different prefix - HDC 43. It was used as a press demonstrator.

10102 - registered LJB 370 and used extensively for early road test reports by the motoring press. Crashed by Roy Brocklehurst during testing in 1959, it was rebodied as a black coupé and used as a back-up in *autobahn* testing in Germany of the B-series engine for the MGB in the autumn of 1961. This and the following cars were all finished in Tyrolite Green.

10103 - registered LRX 238 and allocated as a demonstrator; sold by the works in 1957. Under restoration in the UK at the time of writing.

10104 - probably registered KMO 326 and used as a works car. This registration was on the car used in the BMC high-speed runs on the Montlhéry circuit, 15 miles south of Paris in October 1955. An MGA, Austin-Healey, Riley Pathfinder, Austin A90 and a Wolseley 6/90 all did 100 miles or more in an hour. The MGA was a carefully-tuned production car which, in driving rain with the hood up, did 102.54mph. Later, with a 3.7 axle ratio, bumpers and hood removed, windscreen replaced by aero-screen and undertray fitted, John Gott did 112.36mph, with a fastest lap of 114.3mph. The car was used rather more slowly by the MGB's designer, Don Hayter, as honeymoon transport. The car was sold by the works, but the registration number was retained and used on the prototype 1600 coupé, which was then used in 10,000 mile autobahn testing in 1961.

10105 - registered MBL 367, known as "Mabel" and used extensively by the competitions department in many European races and rallies. At the time of writing owned by MG expert, Bob West, in the UK who has restored it to Mille Miglia specification.

came off what was called the finishing line on July 27, suggesting that some development was done after it was built. This historic MGA, the first of 1000, was a right-hand drive car, finished in Tyrolite Green with grey upholstery and registered LJB 205.

At the same time, the left-hand drive prototype, to North American specification, was built in the competitions department. It was Orient Red with black trim and had the same chassis number, but a different prefix - HDC 43.

Chassis 10102 to 10105 - all finished in Tyrolite Green - were built before the end of May and allocated to various factory departments. Most of their history is known to the MG Car Club. (See panel).

The colours for the new car had been chosen by Syd Enever and batches of five cars were run off in each colour to see what they looked like. After the initial five in Tyrolite Green, five were finished in Orient Red with black trim and the same number in Glacier Blue with grey trim, White with black trim and the last batch, from chassis 10121, were Black with red and black trim. All were destined for export. Chassis 10106, finished in red, was the first to be despatched from the factory on July 13 1955, destined for North America.

The lines began slowly to gather speed as the summer of 1955 ended. Six cars a day were being produced by October, 18 a day by November and 27 by December. A stream of cars with left-hand drive specification trickled out of the factory, destined mainly for North America. The first for the British market were demonstrators for MG's London distributor, University Motors, chassis 10207/8 and 10210/1 built in September.

The Motor Show at which the MGA made its first public appearance was

One of the first publicity pictures of a home market car, taken on the turntable at Cowley.

The ventilated disc wheels suited the car well. Later 1500s had lighter and stronger wheels fitted and the ventilation holes were slightly oval. Some at Abingdon thought the wheels looked so good they could not understand why wires were often specified by customers.

in Frankfurt in September 1955. Rather than being transported by land, the car was flown there in a Lockheed Constellation.

The Earls Court Motor Show cars, which followed, caused immense interest, and were built at the end of September. They appeared, fittingly enough, in red, white and blue - the latter in left-hand drive form. The brochure accompanying them talked of a car with a maximum speed of 95mph which could cruise all day at 80mph.

By the time the year ended, total output had reached 1063 cars and all but 90 of these were for export, with specially-prepared versions being built for motor shows all over the world - Copenhagen, Amsterdam, Montevideo - interest in the new car was enormous. By May of the following year, the factory was expanded to produce extra output. Production had increased by a third since January and at the New York Motor Show 3600 orders were taken for the MGA.

BMC confidently predicted that they would soon be exporting 12 to 15 thousand sports cars (MGAs and Austin-Healeys) a year to North America. In the event, in the following year nearly 14,000 MGAs alone were destined for North America.

"A jolly good little sports car"

One of the curious things about the official road tests which followed the car's production was that very few reviews, other than some in America, said what a very much better car the MGA was than its predecessor. Although they were both 1.5-litre sports cars from the same factory, very few comparisons were drawn; it was if the TF was an embarrassing elderly relative who had outlived his welcome and was not mentioned in polite company. Qualitative comparisons were hard to find and this could be because hardly any of the last version of the TF, the 1500, were sold in the UK and the car was not tested by the British magazines. The *Autocar* did note that "... the new car appears to be a worthy successor to the famous and well-loved T-types, incorporating many lessons learned in racing". And *Motor* commented: "The title MG Midget, which has been associated with a series of four-cylinder two-seat models since 1929.....is now in abeyance." It described its wind-cheating body as being a complete departure from the shape of previous models and pointed out that at 60mph, it required 27 per cent less power to maintain its speed. Off with the old, welcome to the new ...

Most road testers of the fifties rarely had a bad word to say in newspapers and magazines about a British product - commenting on the size of the ashtray was considered blistering criticism by some and there were notable sins of omission when it came to writing of reliability and build quality. But there was no need for reticence when the major magazines tested the MGA, since they all applauded its looks, speed and handling. And the verdict of history is that they were right on these matters; the other criticisms were minor.

One of the most enthusiastic testers was John Bolster in *Autosport*, who had the reputation of being the most outspoken of the motoring writers. He reminded his readers that he had driven the Le Mans car from which, he said, the MGA was derived. This was not really true, since the first road cars were already being produced in June 1955 when the Le Mans cars were prepared. Nevertheless, associa-

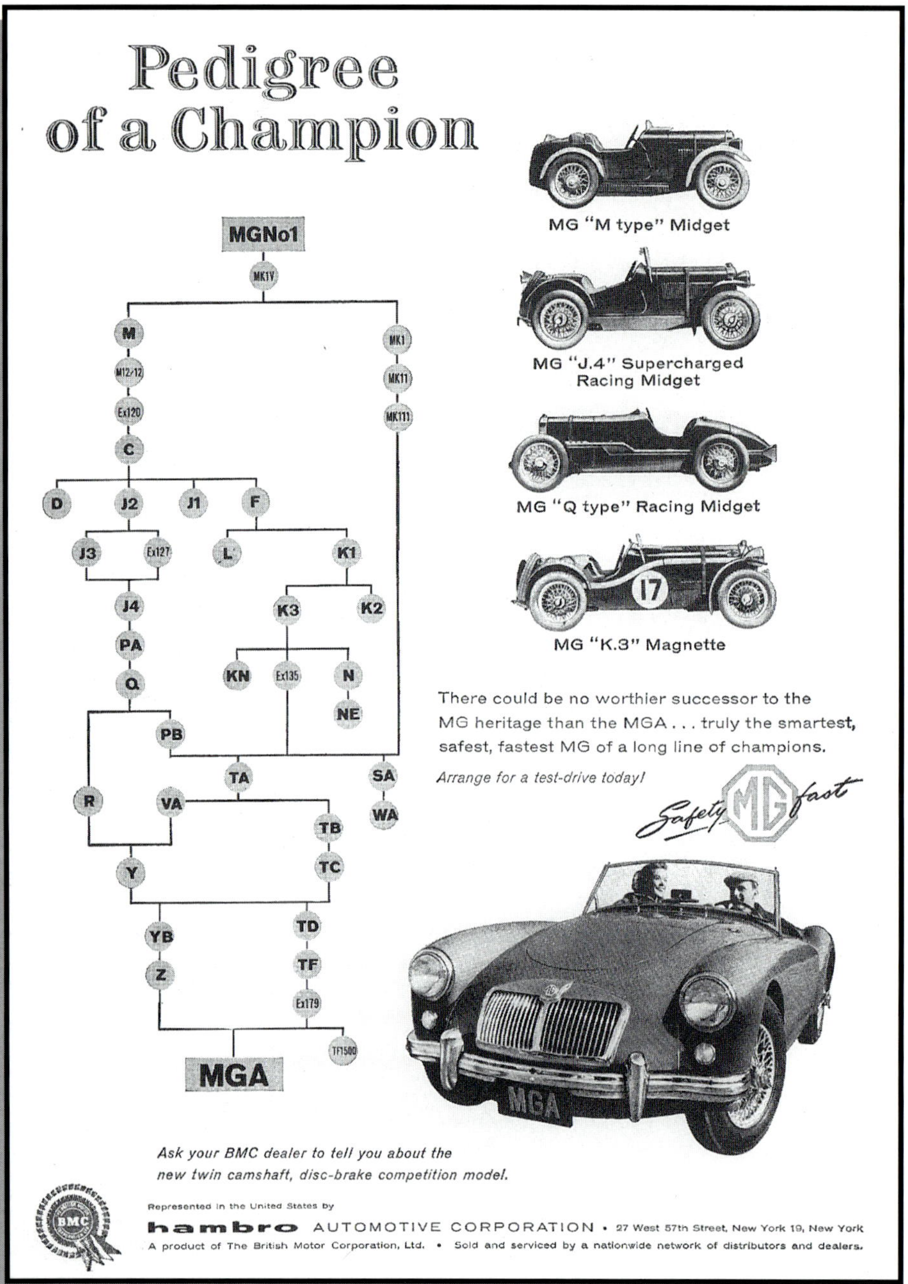

Publicity - particularly in the US - made much of the MGA's sporting heritage.

tion with a racing heritage was helpful.

Bolster went on: "I can say that little has been lost and a great deal gained in grooming the machine for production. The excellent roadholding and steering of the prototype are fully retained and the loss in performance is less than I expected. This is a jolly good little sports car; if you want one, hurry up and get in the queue".

He found it fast and a delight to drive, as well as being comfortable and practical. Not only that, it had excited admiration wherever he went and he claimed that the more discerning were quick to tell him that it was beautifully made. He was particularly impressed with the gearbox. "It is as nearly crash-proof as anything I have driven outside the automatic class ... after being balked by a slow vehicle one takes a *coup de troisiéme* and the speedom-

What SCH Davis thought

One of the first journalists to test the new car was the famous motoring scribe SCH (Sammy) Davis. He had the chance to drive it in August 1955 and gave John Thornley his views. Writing from his home in Woking, Surrey, to John , he gave an unvarnished view of the MGA, quite unlike the rather bland comments which were to appear in the motoring magazines:

Just a line about the new car. Without any doubt it is better than its predecessors especially in speed, cornering and brakes. Would be interested to know how fast it could really take a curve.

Don't like the position of the steering wheel one little bit and am enraged by an old fashioned jack. You want a drill for putting hood up single handed but it is a better hood so are the side curtains. Ventilation might be looked to, cockpit was getting hot.

Suspension a bit hard at speeds up to 40. Would like Healey pattern seat. Throttle pedal should be altered to allow heel and toe.

Like the way the engine auxilliaries [sic] *are accessible in that small space. Didn't check where the fuel pump is, hope not underneath.* [it was].

Like Healey alluminium [sic] *edging to cockpit better than leather. Would like Healey ash tray. Performance excellent for money don't know what 6000 is on top but got it by mistake owing to odd marking of rev counter.*

All the best
Yours
Sammy

A better sports car

So, how much better was the MGA than the TF which preceded it?

		MGA	TF 1250	TF 1500
Capacity		1489cc	1250cc	1466cc
Max power		68 bhp* @ 5500rpm	57.5bhp @ 5500rpm	63 bhp @ 5000rpm
Max torque		77.4lb ft @ 3500rpm	65lb ft @ 3000rpm	76lb ft @ 3000rpm
0-60mph		15.6secs	18.9secs	16.3secs
Top speed		98 mph	82mph	85mph
Weight		1988lbs**	1930lbs	1932lbs
Length		13ft	12ft 3ins	12ft 3ins
Height		4ft 2ins	4ft 4.5ins	4ft 4.5ins
Width		4ft 10ins	4ft 11$^{3/4}$ins	4ft 11$^{3/4}$ins
Wheelbase		7ft 10ins	7ft 10ins	7ft 10ins
Track	*front* *rear*	3ft 11$^{1/2}$ins 4ft $^{3/4}$ins	3ft 11$^{3/8}$ins 4ft 2ins	3ft 11$^{3/8}$ins 4ft 2ins
Cost (1955)		£844	£780	not catalogued in UK

**At launch; later 72bhp.*
*** Manufacturer's kerb weight figure*

MGA figures from *Autocar*; TF figures are taken from *Road and Track*, as no UK magazine tested the TF 1500.

Comparisons with the earlier TD and TA are more marked. It was notable that when *Motor* compared the MGA with the two earlier cars, it found that there was little difference between the 1937 TA and the 1950 TD. But the leap in performance between the TD and the MGA was substantial.

	Weight (lbs)	Max mph	0-50mph	mpg	Cost @ 1957 prices
1937 TA	2139	77	15.2	32	£666
1950 TD	2408	77	13.5	26.3	£670
1956 MGA	2015*	98	10.8	26.7	£595

**Dependent on equipment. Tested weights ranged between 1988lbs and 2436lbs.*

The figures emphasised how out-of-date the TD was in 1950, as it was effectively a 1930's car. The MGA was 25 per cent faster through the gears to 50mph, 28 per cent faster at the maximum - and cheaper!

eter is soon climbing into the seventies again". They don't write like that any more.

Handling was obviously something on which the testers concentrated. *Motor* thought the car an enthusiast's delight, with light steering and cornering which inspired complete confidence. The magazine recommended

Left: The early instruments received a poor reception and were soon changed as they were considered not precise enough. Pictured is the dash of John Eagle's car, built in November 1955 at a time when facias of black or green cars were painted red, rather than the body colour.

Later cars had much more precise ladder scales. The rev counter in this car has been changed to an electronic type and is a Smiths rather than a Jaeger, though both makes were manufactured by Smiths.

that the high-speed tyre pressures of 18lb at the front and 23lb at the rear should be increased to 26lb all round, which must have made the front end float about a bit.

Autocar reported that fast cornering was a joy and long winding hillside roads a joy to traverse. This joyous note continued with gearchanging and braking being praised: the Lockheed hydraulic drum brakes were widely appreciated.

As for interior comfort and equipment, John Bolster felt the steering wheel should be further from the driver (he didn't have the telescopic option in the test car) and he wanted the seat cushions improved. The *Autocar* also wanted more support for thighs, but there was praise from *Motor* for the comfortable seats: "well sprung, well covered in leather and excellently shaped to prevent passengers being flung around in fast cornering". But *Motor* was not keen on the instrument dials, which it said had suffered at the hands of the stylist. This was a reference to the very large figures scattered around the dials at wide intervals with no ladder scales between them. (These dials, made by Smiths, but marked Jaeger, were soon changed to more traditional instruments).

The *Motor* tester appeared to have some difficulty with the hood: "... to raise or even lower the hood single-handed is an exercise in skill and patience." he said tactfully. The *Autocar* pointed out rather obviously that it was easier for two people to raise it than one and said there was a reasonable amount of headroom with the hood up, because its driver didn't hit his head on the hoop sticks whilst going over bumps.

The door and window arrangements were primitive, though no-one said so. The detachable sidescreens had a spring-loaded signalling flap - replaced in later models by sliding sidescreens - and the door latch was

The first tests

	Motor LJB 439	**Autocar LJB 370**	**Autosport LJB 370**	**Road & Track**
0-60mph	16secs	15.6secs	15secs	14.5secs
Max mph	97.8mph	98mph	96.7mph	97.5mph
Fuel	26.7mpg	27mpg	29mpg	28/32mpg (US)

Hambro, the US MG importer, advertised the new car heavily, and meeting demand was often a problem.

operated by a pull cable in the door pocket. It seems strange in these security-conscious days that there were no external door handles and the car's doors could not be locked. The boot was released by a handle behind the passenger seat. However, it was common not to have exterior door handles in sports cars. Although the T-series Midgets had them, the Jaguar XK 120 and 140 and the TR2 and TR3 did not and the reason had as much to do with economy as it did aesthetics.

Reaction to the new car in North America was so favourable that the importers, Hambro Automotive Corporation, published a booklet in July 1956 containing the first road tests from *Road & Track, Motor Life, Mechanix Illustrated and Motorsport.* They testified, said Hambro proudly, to the "outstanding styling, roadability and nimble performance of the new MGA ... truly the smartest, safest, fastest MG ever engineered!" The booklet did not include the test from *Motor Trend* in which one Walt Woron talked of pleasurable power pulses from the "violently-running four", but wanted more power. "Why saddle it with a 1.5-liter engine that doesn't even allow it to hold its own against the '55 Chevy, Ford or Plymouth when leaving the lights?"

Road & Track had no such reservations; early enthusiasm for the car's appearance, it said, was augmented by the knowledge of its very surprising performance. The magazine praised its ride and cornering - the perfect car for the tyro enthusiast. There was lavish and unexpected praise for the side curtains - "a touch of genius" - and the weather sealing was so tight that "some ventilation ought to be provided to prevent suffocation". Latter-day owners may beg to differ!

The *Mechanix Ilustrated* test was done in the UK by Gordon Wilkins in LJB 439, who concluded: "The fact remains that it is more than ever fun on four wheels at a price that the average fan can pay".

Motor Life was one of the few magazines that compared the MGA to the TF. "Completely new, better than the T-series and entirely different". The other was *Motorsport*: "gone is that nervously high-strung feeling over 60mph". Handling, it said, was of the very best - so was the workmanship. Speed and power were more than adequate, but not sensational compared with the German Porsche Super and

Jack Hawkins takes a break from filming The Man in the Sky *to pose with an early 1500.*

the British Triumph TR2.

All in all, wherever it was tested, the verdict on the new MGA was immensely favourable. As *Motor* said, the famous MG slogan "Safety Fast" was never more aptly applied.

Production picked up to 13,410 cars in 1956 and 20,571 in 1957, when the roadster went up £35 to £995 including tax, and the new coupé rose by £37 to £1087. Production dropped back to 16,663 in 1958 and 7644 in 1959 as the Twin Cam and new 1600 came on stream.

The last 1500, a left-hand drive car in Island Green, chassis 68850, was completed on June 3 1959, while production of the 1600 was already going ahead.

The MGA 1500 was a great success, cornering the market in its class. Its only competitors, said *Autocar* in May 1957, were the Volkswagen Karmann Ghia and the Simca Weekend and Oceane. (Whatever happened to these last two?) This statement was more than passing strange, since MG believed the car's chief competitor to be the TR3 and said so in notes marked "Confidential" which were circulated to sales staff, defining the MGA roadster in an excruciating cliché as "the car for the young executive or the executive who is young at heart". This was accompanied by a drawing of the mythical figure - a serious chap with pipe and glasses. If this mythical potential customer were to point out that performance of other sports cars was better, the sales staff had to reply that the MGA's was a SAFE (in capital letters) performance, thrilling enough to satisfy most enthusiasts. This was perhaps to make the most of the TR's reputation of having unusual handling characteristics.

The secret briefing, headlined "It's their sort of car - sell it to them now", conceded that the TR3, with its larger engine of 1991cc, was able to return what was called "slightly better figures", especially with overdrive fitted. (The MGA was never offered with overdrive, though some engineering drawings were done in May 1957). In fact, 0-80mph in the TR took 20.7secs compared with the MGA's 32.1secs, and the top speed of the Triumph was 107mph, whilst the 1500 roadster could not reach the ton.

Sales staff were told to emphasize the MGA's other plus points: leather upholstery instead of the TR's leathercloth; a turning circle of 29.25ft against 33ft; the optional heater being the fresh-air type rather than a recirculatory unit, and an inch more headroom.

There was, however, no mention of brakes; in 1956 the TR3 was the world's first series production car with disc brakes, which the pushrod MGA did not acquire until the 1600 model in 1959.

Competition for the MGA's market at home was limited, unlike North America where the potential market was enormous and disposable income far greater. Italian and German exotica, such as the Alfa Romeo Giulietta Spyder and Porsche 356 have been mentioned as rivals in the UK, but these cars were far more expensive and lacked the huge dealer network that BMC had. The Alfa, for example, cost £2248 in 1958 - about double the price of a Twin Cam. The Porsche 1600 convertible cost marginally under £2000, as did the home-grown Lotus Elite.

The MGA's like-for-like rivals were the Morgan Plus Four, built in tiny numbers and retailing in 1958 at £968 and, from the following year, the Sunbeam Alpine - £971 at that year's London Motor Show, compared to the MGA 1600 at £940. The Alpine was to blossom briefly, then fall prey to the MGA's successor.

Making the MGA

As 1955 ended, the two production lines continued to gather speed and, some 18 months after manufacture began, nearly 100 MGAs a day were being made. Production of the MGA

The cost of building an MGA

Despite the immediate success of the MGA in both home and overseas markets, it rarely made a profit for Abingdon and sometimes made a loss. This was due in part to the arcane accounting practices of the BMC Group, but also because overseas distributors' discount was considerably higher than for UK dealers: in the United States, they got no less than 33.3 per cent

In early 1957, every MG sold in North America was sold at a loss, and not only the MGA; the Magnette, recently exported there, was losing £79 a car.

These costings for the MGA in March 1957 might be of interest:

Material		£383 10s 8d
*Less 3/43rds of £199 2s 3d**		£13 17s 10d
		£369 12s 10d
Labour		£10 7s 9d
Overheads at 170 per cent		£17 13s 0d
		£397 13s 7d
Publicity		£25 0s 0d
Free service		£3 0s 0d
		£425 13s 7d
Group profit		£79 18s 5d
Receivable		£505 12s 0d
Distributors' discount 21 per cent		£134 8s 0d
Home retail	£640 0s 0d	
Purchase tax	£321 7s 0d	
Selling price	£961 7s 0d	

** It's not known what this figure represents, but it is to do with inter-plant accounts at BMC.*

John Thornley, an accountant himself, railed against what he saw as BMC's crazy financial policies and inter-plant costings. For instance, if the MGA costings were done the same way as at Longbridge, the selling price would be not be £961 but £809. Another anomaly was that the US distributors' discount on an Austin-Healey was 21.5 per cent rather than MG's 33.3 per cent.

Despite record output and stable labour and overhead costs, the factory showed a loss in many of its trading periods and a minimal profit in others. In the 36 weeks to April 7 1956, the period covering the introduction of the MGA, the MG Car Company showed a profit of just £225.

BMC's financial staff, said Thornley, were like Hannah in the song 'pouring water on a drowning man'. The official position was rather different. When *Motor Sport* visited the factory a year later, when 99 per cent of MGA production was going abroad, it was stated that the staff of 1030 at Abingdon were earning $500 per operative per week in export sales, such was their importance to the UK export drive.

Saving the pennies

BMC, like most large manufacturers, always built down to a price, paring fractions of pennies off the cost of mundane items. The original door latch for the MGA was costed by the makers Wilmot Breeden at 4/1d. It was rejected in favour of one costing $3/11^{1/2}$d. Multiply the difference by 100,000 cars and the saving was more than the basic price of one whole car. Adding external handles would have cost as much as several more cars.

This parsimony was noted by *Motor* among others, who commented that the virtues which gave the MGA its personality had been acquired at the expense of creature comforts: although there were two map pockets, there were no glove compartments, parcel shelves, sun vizors, ashtrays or cigar lighters. On the other hand, there was a generous tool kit stowed in the boot, though its comprehensive nature suggested that either trouble or extensive home maintenance was expected! It comprised a screw-type jack, wheelbrace (or copper hammer for wire wheels), tappet spanner, tappet gauge, spark plug spanner, two tyre levers, cylinder head nut spanner, tyre valve spanner, distributor spanners, three open-ended spanners, screwdriver, recessed screwdriver, tommy bar, brake bleeder tube, gearbox plug spanner and a touch-up paint pencil.

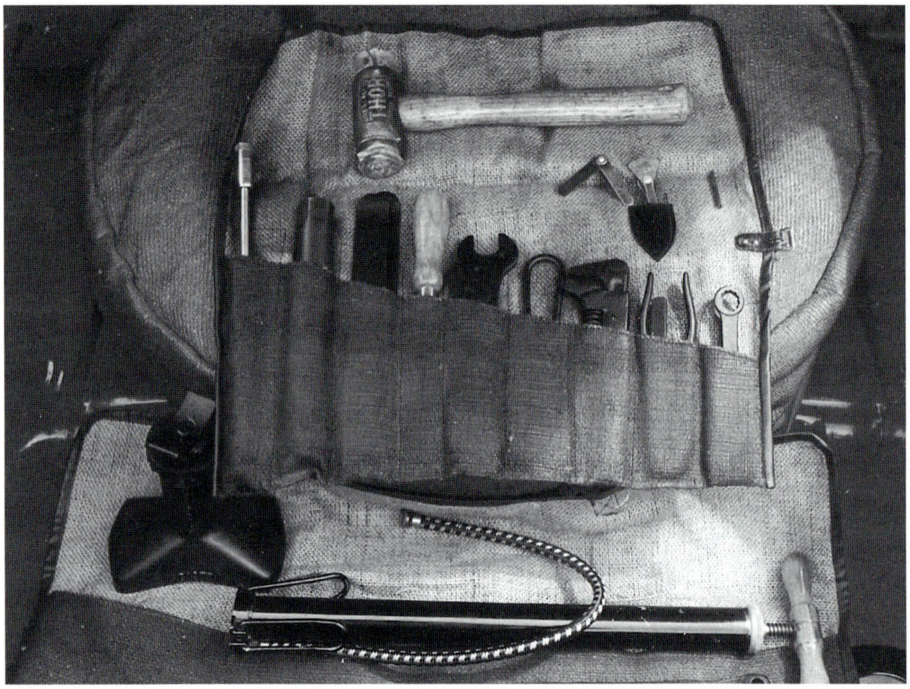

The toolkit was lavish by today's standards. This is an original kit in the boot of Ian Bowen's 1959 coupé. In later coupés there was even less room in the boot.

marked a major change for MG, not just in terms of the car's structure, but also because it made Abingdon a mass-producer of sports cars rather than a small specialist manufacturer.

In 1954, the last year of T-series production, 6520 TFs were made. Five years later, in 1959, Abingdon was to turn out more than 23,000 MGAs in a year, plus Austin-Healeys and oddities like Morris Minor variants, produced with a staff of just over 1000. John Thornley used to point out that if you took away the 300 or so people needed for administration, the output of cars per worker was phenomenally high.

This experience of mass-production with the MGA was to lead to huge volumes of 30,000 and 40,000 MG sports cars a year being produced in the late sixties and early seventies. The remarkable factor in this shift to mass-production was that the cars were constructed in exactly the same way from the day the factory opened until the day it closed. There were no big overhead conveyors or powered tracks; chassis were pushed along by hand, having components added until they reached the end of the production line.

Abingdon was a tiny factory compared to the Longbridge and Cowley leviathans, there was little room for expansion and hardly any components were made there. High MGA production was achieved by assembling the major components off-site and uniting them on the track. The Morris Motors bodies branch made the shell from the panels manufactured by Pressed Steel. This was a substantial construction, delivered to Abingdon with all the main panels welded in place. Wings came separately, on occasion, to be bolted on and were finished by panel beaters and sprayed before attachment. Doors, bonnet and boot lid, added later, were skinned with aluminium.

The bodies arrived from Coventry, painted and trimmed, in big open double-decker transporters, three up and three down. The trucks waited in a marshalling area until a series of coloured lights signalled them to unload on the first floor - 'top deck' - as it was known. The bodies were mounted on four-wheel trolleys and pushed along the floor as Lucas electrical equipment and other components were added.

On the ground floor below, the chassis was taking shape in the press shop. It was built up from box-section sidemembers, tubular crossmembers and the most-important 'goal post' structure which kept the scuttle rigid.

The components came from John Thompson of Wolverhampton and were largely of heavy 14-gauge steel. They were put first in a tacking jig to ensure they were properly assembled, before being transferred to a welding jig, which could be rotated to allow the two-man welding team to reach all parts of the frame. Any slag left on the welds was chipped away with hammers. Each chassis was assembled by a pair of men, who could produce 20 frames a day. As a quality control, one frame in every hundred was tested for accuracy.

When complete, the chassis was dipped in black stove enamel, allowed to drip and then dried for six minutes in a gas-fired, infra-red oven.

Front and rear suspension pressings were added next, with wheels and tyres; tyre fitters like Bernard Snuggs and Phil Boyce could fit 400 tyres a day, using one small lever. Dunlop or Michelin tyres were usually specified.

The Austin-Healey chassis then began its progress down the production line, pushed by hand along a pair of raised tracks made of concrete blocks with a metal channel in which the left-hand wheels could run while the right-hand wheels were on a flat surface. In this way the lines could accommodate cars of different track widths.

The almost-complete body was lowered from the first floor on to the waiting chassis by a Feldo electric crane. (Assembly differed to that of the Austin-Healey 100-6 which was produced on parallel lines at Abingdon.

Built to order

MGA brochures and road test reports always featured the list of extras which were available with the car, though usually this covered only a few options, since it was possible to get a car built to order with many non-standard features.

An example was the covering on the facia, which could be Rexine, vinyl or leather and in whatever colour was required.

Body colours were also available to special order. British Racing Green was very rarely ordered; it became popular later on the MGB. Another rare colour was Pearl White - developed by Syd Enever with paint-makers I.C.I. - which contained ground mother-of-pearl. Enever's own MGA, which his son subsequently acquired, was finished in this colour. The most lurid was Pale Gold Metalochrome, similar to that used on the 100,000th MGA. Many models were made for prominent people, among them BMC's chairman Sir Len Lord and King Hussein of Jordan.

Heaters did not become standard on MGs until the late sixties, so warmth from a recirculating or fresh air heater cost extra. So did a radio, wire wheels, fog lamp, whitewall tyres, 4.55 to 1 rear axle instead of the standard 4.3:1, external luggage carrier on the boot, tonneau cover, radiator blind, rim embellishers and a telescopic steering column. Some of the rarer options available were: Ace Mercury wheel discs; headlamp flasher relay and switch; sunvisors (only on the coupé). One of the most popular accessories, which could be fitted by the factory or the dealer, was the boot-mounted luggage grid. Given the shortage of space in the boot, particularly in the later coupés, this was not surprising.

Also on offer, from the Automobile and Sheet Metal company from January 1958, was a grid which attached to the rear bumper fittings and hinged outwards and backwards, so that it lay parallel to the road, overhanging the back of the car. This required a 9ins extension to the tailpipe to avoid luggage becoming blackened.

Export models were offered with what was misleadingly called a Cold Air Ventilation Kit, which implied air conditioning but was simply a heater kit without the part that produced the heat. The air was as cold as the ambient temperature outside the car which, in many overseas areas, would be rather hot. For colder climes there were radiator blinds and muffs.

There were useful things available like wing mirrors and fog lamps; dealers looking to boost their mark-up were pleased to fit approved trinkets like badge bars and monogrammed exhaust deflectors.

The list went on and on, most of it containing items which are fitted as standard in today's cars - ashtrays and cigarette lighters, for example. There was also a wide range of mechanical enhancements available, which are covered in the chapter on tuning.

Overdrive was never offered as an option on the MGA although it was thought about in some detail in the spring of 1957 when drawings were done of the changes that would be necessary to accommodate it. These involved alterations to the crossmember, floor supports, the gearbox cover and the remote control cover. It seems likely that overdrive was not adopted because of the cost involved in making the changes. Further experiments were done in July 1959 with a Laycock overdrive, probably in preparation for the MGB, which was to offer overdrive and automatic as options.

The body/chassis arrived fully wired with instruments in place and partly-trimmed).

The B-series engine, run-in by electric motors at Longbridge with the sump off, was dropped in from an Ardegar hoist. Manifolds were added with the SU carburettors, engine mounting rubbers and the special remote-control casting for the gearbox - a bought-in part not specified for any other BMC car.

Each car coming down the line had a note on the windscreen specifying the type of wheels, lamps and speedometer needed and whether the car was to be fitted with a left- or right-hand Cam Gears rack and pinion system, which it inherited from the TF. Specification was decided by the planning department, who analysed customer orders. White was the most popular colour in the late fifties and, because of variations in the fitting of steering, heaters, wheels and radios, only about two cars in every 500 were exactly the same.

In 1957, 60 per cent of customers specified white as a body colour and the same number wanted wire wheels, a thirties anachronism disliked by the factory management, who thought their pierced wheels elegant enough, though the first production car had wire wheels.

Not every car was as it should be. Henry Stone recalled that in his time at Abingdon he had seen a right-hand drive car coming off the line with the pedals on the left and another with wire wheels on the front and disc wheels on the back.

Each operation on the line took about eight minutes and, at the end, lights were checked and steering tracked on a Weaver machine to measure the toe-in. Inspectors checked the cars as they neared the end of the line. As they were on a flat-rate wage rather

The B-series engine is shoe-horned into the chassis with a hoist.

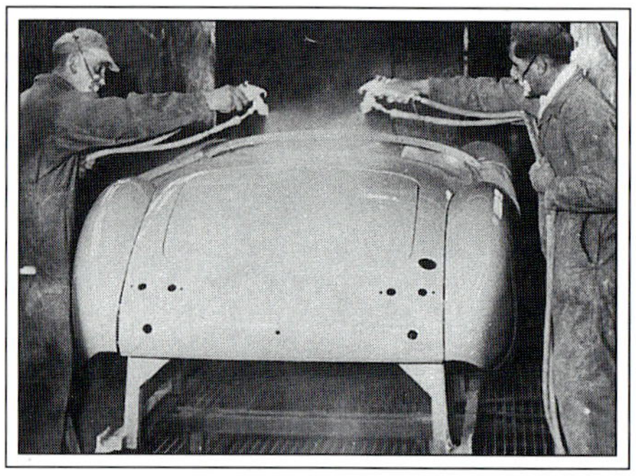

The body was sprayed by hand and delivered to the top floor of the Abingdon factory where electrics and other components were fitted.

than the piece-work rates of the assembly-line workers, the inspectors took good care to check that assembly had been carried out correctly.

Test drivers then took each car on a six-mile road test, an MG tradition which persisted up until the seventies. It was a common sight in Abingdon to see MGs on trade plates sweeping out

Right: The body was lowered on to the chassis and the connections made while the bonnet was held open with a special prop.

Left: Nearing the end of the production line in 1956. The ticket on the windscreen gave details of specification and destination.

of the Marcham Road gate, down the A415, up to the Sheepstead crossroads and back to the factory. Drivers noted any faults in a log book which comprised the body and chassis cards stapled together - main MGA troubles were loose suspension mountings and flawed paintwork - and passed the cars to the rectification department. Faults

It would never look this clean again .... final inspection before the 6-mile road test.

Below: The October 1956 launch of the coupé brought comforts like wind-up windows to the formerly spartan MG sports car range.

in the finish were marked with chalk rings and retouched or polished out under neon striplights. After a final wash and polish the vehicles were passed fit to leave the works.

About 15 cars an hour were going through this process in 1957, according to *Motor Sport,* which visited the works to chronicle the MGA's progress. Cars destined for export - which meant most of them - were covered with protective wax or 'export jam', as it was known.

In those days many keen British MGA buyers picked up their new cars from Abingdon and were allowed four gallons of premium petrol from the factory pump. But the great majority of cars, those for export, were ferried by teams of drivers - mainly women who returned by bus to pick up more cars - to the marshalling bays 13 miles away at Cowley. The MGAs were loaded on to freight wagons and transported by rail to the docks for their journey across the world.

The coupé

It has long been a ploy of the motor manufacturers to maintain interest in a model for several years by bringing out carefully-staggered variations. So it was with the MGA, for a year after the launch of the roadster the neat coupé appeared, developed to meet the demand for comfortable fast cars, the new GT genre, which provided the excitement of speed without the inconveniences of the open tourer. Alfa Romeo and Jaguar led the way with stylish coupé bodies.

The coupé was often described in

Left: The neat pull-down door handle was unusual in style.

Right: The hard top accentuated the rear haunches of the car. The dome was welded in place at Morris Bodies.

magazine price lists as a "hardtop", though that name is more properly applied to the after-market removable tops which were quickly designed for the convertible MGA. The first, a detachable metal top, appeared in January 1956, marketed by GBG Accessories of London. This was followed a few months later by an offering from Universal Laminations; a quarter-inch thick glassfibre affair, ready-wired for a roof lamp. The factory version was not ready until the spring of 1956.

Two development cars - both red - were built in December 1955. 10869 and 10870, one with home market specification and the other with North American, were noted in the records as "hardtop development". One of the cars was sent to coachbuilder Vanden Plas, part of BMC since takeover by Austin in the forties. Vanden Plas developed a black aluminium top with a wrap-around one-piece rear screen and this became available from the summer of 1956. It had sliding windows in aluminium frames and was used on the MGAs in the Alpine Rally of that year. The rear was fixed by two tongues which engaged in slots in the rear deck behind the seats, and secured in the front by wing nuts.

Jim O'Neill, MG's body man, later designed a different hardtop, which was made by Universal Laminations in heavy glassfibre. Again, it had sliding sidescreens. It was available in vinyl-covered black glassfibre only. In addition to this official factory version, there were also tops available made by component manufacturers.

The job of designing the coupé proper was given to the versatile Eric Carter at Morris Motors bodies branch. In March 1956 he took a black roadster, chassis 13355, off the line to use as a development car.

The news reached Abingdon that the new top was ready to be viewed and approved. John Thornley, Syd Enever and Jim O'Neill went to Coventry to see it. O'Neill was in for a shock: "I was horrified", he said, "the top looked like a pimple on a haystack - quite wrong. I would have made it longer, a more elongated design. "Eric said to me 'don't say too much, because we have already started tooling without permission and if we stop now all that tooling will have gone to waste'. "So I didn't say anything and now, of course, it's a classic shape. In retrospect Eric did a really smashing job - it's a classic motor car". (The car - ULJ 426 - still exists). It was the first MG coupé since the thirties (and most of those were built by specialist coachbuilders rather than by the factory).

Little attempt was made with the new coupé to get away from the basic shape of the MGA, either with hood raised or removable hardtop. Compared to the MGB GT, which put a "shed", as John Thornley called it, on the MGB roadster, little body engineering to the basic shell was required, other than to slightly alter the immediate areas where the top met the body, and adopt door skins with a different profile.

The top was a very clever one-piece steel pressing with a reinforcing panel inside and special windscreen pillars. The windscreen had a deeper curve than that of the roadster, but what made the car really distinctive was the splendid wrap-around rear window, whose curvature was made possible by dividing it into a three-piece unit - one main piece of glass and two side pieces, jointed by thin rubber pillars.

There was a deliberate attempt to make it more comfortable than the roadster; it had wind-up windows, different seat trim, sound-deadening foam behind the Vynide-covered facia, and a carpeted space behind the seats. The loss of pockets in the doors caused by the window mechanisms was remedied by supplying map pockets in the footwells. Not only was the car lockable (unlike any other MGA), it also had ingenious vertical door handles, almost imperceptible from a distance, operated by being pulled downwards, a great deal more sophisticated than the internal cord pull used in the roadster.

The first three cars came off the production line in the second week of October 1956. Car/chassis 20671, red with black upholstery, went to University Motors, MG's London distributor; 20672, in white, was the London Motor Show car and 20673, green with grey upholstery and with a kilometre speedometer, was built to North American specification and displayed at Nuffield House, BMC's West End showrooms.

The coupé was designed to appeal to women as well as the traditional male buyer.

Though on show, examples were thin on the ground. The first press demonstrator (21210) was not built until the middle of November. By the end of 1956, only 16 had been built and the first major batch for North America, comprising 12 coupés, was not despatched until February of the following year.

The coupé was originally catalogued in the UK at £1049, which was £88 dearer than the roadster, though purchase tax variations distorted the basic cost, which was £640 for the roadster and £699 for the coupé - a difference of only £59. In the US the basic list price was $2750. (It is often very difficult to assess the comparative prices of cars in the fifties because of large and frequent variations of the rate of purchase tax, which was used by the Government to try to control the economy.

The new top and wind-up windows imposed a weight penalty. The kerb weight of the roadster with five gallons of fuel was 2009lbs (4419 kilos), according to *Autocar* figures; the coupé, in the same trim, was 2107lbs (4635 kilos) a difference of 98lbs (215 kilos). But the smoother profile reduced drag and, for the first time in the company's history, MG had a genuine 100mph production sports car. It was some three to four miles an hour faster than the roadster and faster through the gears - some three seconds quicker to 80mph and five seconds to 90mph.

The sales promotion notes given to the dealers promoted it as a model for customers who liked to motor fast in all kinds of weather: "It gives him the motoring he loves and her the comfort she demands".

The first serious test was done by *Autocar* in April 1957, which said: "An excellent successor to a famous line, with fine roadholding and a performance which is worthy of the manufacturer. Where the going is tricky, it should prove capable of holding its own against much more powerful rivals with larger engines". However, getting in and out was a problem and there was little stowage space.

Most of the road tests did not appear until the summer of 1957. John Bolster in *Autosport* declared it: "a proper little coupé and it will sell like hot cakes ... a genuine closed car with wind-up windows and not an open two-seat with a hardtop". He called the coupé "astonishingly fast", taking it to 102.27mph at which speed the valves began to bounce.

Criticisms were minor: it rolled fractionally more than the roadster on cornering and getting in and out was not particularly easy because of the restricted door opening.

Motor, in August 1957, took the coupé on the banked Montlhéry circuit and discovered it to be "remarkably fast, exceptionally safe and a constant pleasure to drive". The testers got up to 101.2mph in overall lap time and managed an impressive 103.8mph on the flying kilometre. They were impressed also by the fact that at 80mph - a comfortable cruising speed - the piston speed was only slightly in excess of 2500 ft/minute at which the engine was delivering little more than half maximum power.

Autocar praised the outstanding flexibility of the engine and got the car up to 102mph.

In the US *Motor Trend* said bafflingly: "This neat little package is liable to make an enthusiast out of grandma" ...

However, it wasn't all praise; the testers from *Motor Sport* and *Road & Track* also found it difficult to get in and out of, largely because the doors

Shortage of boot space, particularly in the later coupé for which the spare wheel was moved backwards, prompted many owners to fit the optional luggage rack.

The first coupé ever built, the one which persuaded MG management to take it into production. In need of some restoration ...

did not open wide enough. *Road & Track*, in particular, was cross about the lack of home comforts, though most of these were available as extras, and wrote: "It is hard to understand why a car aimed at Americans (as well as others, naturally) should lack the minor conveniences that we softies have come to value, but lack them it does: there is no ashtray, no cigarette lighter, no sun visors, no side armrests and no glove compartment ... the seats do not give enough support on long trips. Also listed in red ink should be the high temperature around the pedals ..."

This besetting problem, which was never cured, was caused by heat emanating from the engine through the footwells. A great benefit to the roadster driver *sans* heater in the winter, but a trial to the coupé owner in the summer.

Noise was another cause for complaint, though this was ameliorated in the 1600 in 1959, when the spare wheel, which had protruded into the cabin in a carpet bag, was moved backwards to be fully enclosed in the boot, allowing a solid bulkhead to be adopted. At the same time, the shelf underneath the rear window was cut back several inches to allow more room behind the seats. Space for luggage, however, was still vestigial.

The prototype car used for this interior revision had a 1500 body with a 1600 engine. It was registered KMO 326, a number which dated back to a 1954 works TF rally car. The car was bronze metallic with a pearlescent white roof. The white paint, which contained real mother-of-pearl, was made by ICI specifically for Syd Enever. The car was thought to have been used extensively by him personally and eventually became a factory development hack, testing the MGB engine.

In September 1961 it was sent to Germany with a 1622cc engine fitted to run 10,000 miles continuously at between 5800 and 6000rpm in top gear. Another MGA, LJB 370, was sent as back-up. This bore the registration

The prototype 1600 coupé, KMO 326, was used extensively as a works test bed for the MGB.

US buyers did not take to the Magnette, and sales of all BMC saloons were a disappointment compared to the roaring success enjoyed by the MGA.

number of the second production MGA, produced in 1955, but is unlikely to have been the same car.

KMO 326 eventually left the factory in 1964, two years after MGA production stopped, and was bought by Syd's son, Roger, and used for rallying. It was given the honorary chassis number of 109071, one more than the last MGA .

The coupé, which ran through all the variants of MGA to the end of production - 1500, 1600, 1600 Mark II, Twin Cam and de Luxe - was never meant to be a high-volume model and, like the MGB GT, which eventually succeeded it, far fewer were built than the tourer from which it sprang.

In the peak production year of 1957, only some 4000 were built, compared with 16,467 roadsters and, by 1960, the proportion reduced even further with 1464 coupés to 15,517 roadsters of all types. And in the last year of MGA production - 1962 - coupé production was down to 115 cars compared to 2944 roadsters.

The discrepancies are not surprising, given that coupé demand from North America was much smaller than for the roadster. Coupé production eventually totalled less than 10 per cent of all MGAs built.

4

THE TWIN CAMS

The strategy

The MGA Twin Cam which appeared in 1958 was originally going to be called the MGB. It is described as such in many of the memos that John Thornley wrote to the BMC board as he pushed for the engine's development, though in later years, after the real MGB had ceased production, he confessed he had completely forgotten the suggestion.

The concept of a competition car had been in his mind for more than a year before the MGA was launched. As far back as August 1954 he had suggested to the board of BMC that Abingdon should produce not just four cars for the following year's Le Mans, but a special series of 50 or 100 competition MGAs. Five or six would be raced by MG and the rest sold to the public.

The aluminium-panelled cars would have twin overhead cam engines, possibly with fuel-injection, five-speed gearboxes or overdrive, and disc brakes. Provided they performed well enough they could be sold readily, recovering development and production costs and even covering some of the racing expenses. Porsche and Osca sold cars in this way for up to £2000 each.

However, whilst Thornley believed the MGA competition car could be sold for £1250 (compared with £844 for the production car at launch), it seems the board of BMC were not convinced and the case for a limited run of competition cars was lost.

A twin cam engine, based on the B-series block, was already available in prototype form in 1954, having been originally created at Morris in Cowley by Gerald Palmer. There were also plans to put his twin cam design on BMC's C-series block, the engine used in the large passenger cars. One engine was built before the plan was dropped.

In 1953, Palmer's remit in the Cowley design office covered both Riley and MG. More than 40 years later he recalled: "It was my opinion that MG ought to have a twin cam motor for better performance, it was high time that they had an up-to-date engine.

"I sketched out the possibility on my drawing board at Cowley then persuaded my boss, Vic Oak, the technical director, to let me go ahead. I then took it to Morris Engines in Coventry for further development work by Eddie Maher, an old Riley man, and Jimmy Thomson. "I left it to Eddie and his lads to decide how it would come out; I can't remember whether the first engine was fuel-injected or had carbs."

The conversion is also believed to have sprung from a performance brief given to Morris Engines to try to run the B-series continuously at 7000rpm on the test bed. Since the standard head restricted breathing at this speed, a twin cam layout seemed the answer since it permitted the gases to get in and out of the engine more efficiently.

Palmer pointed out the other benefits of a twin overhead camshaft: maximum engine power is increased because the camshaft does not have to shift the inertia of pushrods and rockers, and the valves can be opposed to each other rather than in-line, which allows larger diameters to be used,

On the right, the Morris-designed Twin Cam and below the unsuccessful Austin challenger installed briefly in one of the Le Mans cars. The engines are superficially similar, but the included angle (the angle at which the valves are opposed) is much narrower on the Austin engine. (Courtesy BMIHT/Rover Group)

getting the mixture in and out more quickly. The end result is an engine which can run at a higher speed and develops more torque than its pushrod equivalent.

While the B-series had the advantage of being cheap to produce and reliable, and could be tuned to a surprising extent, it was not a competition engine. John Thornley is quoted by Robert Vitrikias as telling him: "We had soldiered on for so long with these silly heads where gas came out the same side that it went in. Moreover, the pushrods were on the same side as the ports - I can't think of anything more ridiculous!"

But not only was a Morris competition engine being developed; Austin at Longbridge had also been working on a prototype twin cam, a completely new unit designed by Bill Appleby which, unlike the Morris version, did not use any existing BMC parts. Palmer said that Appleby's Austin design had been heavily influenced by the air-flow consultant, Harry Weslake, who was doing a great deal of work for BMC at the time, and who had, more significantly, been involved with Jaguar in the engineering of the XK 120 with its double overhead camshaft engine.

"I wanted my twin cam to have

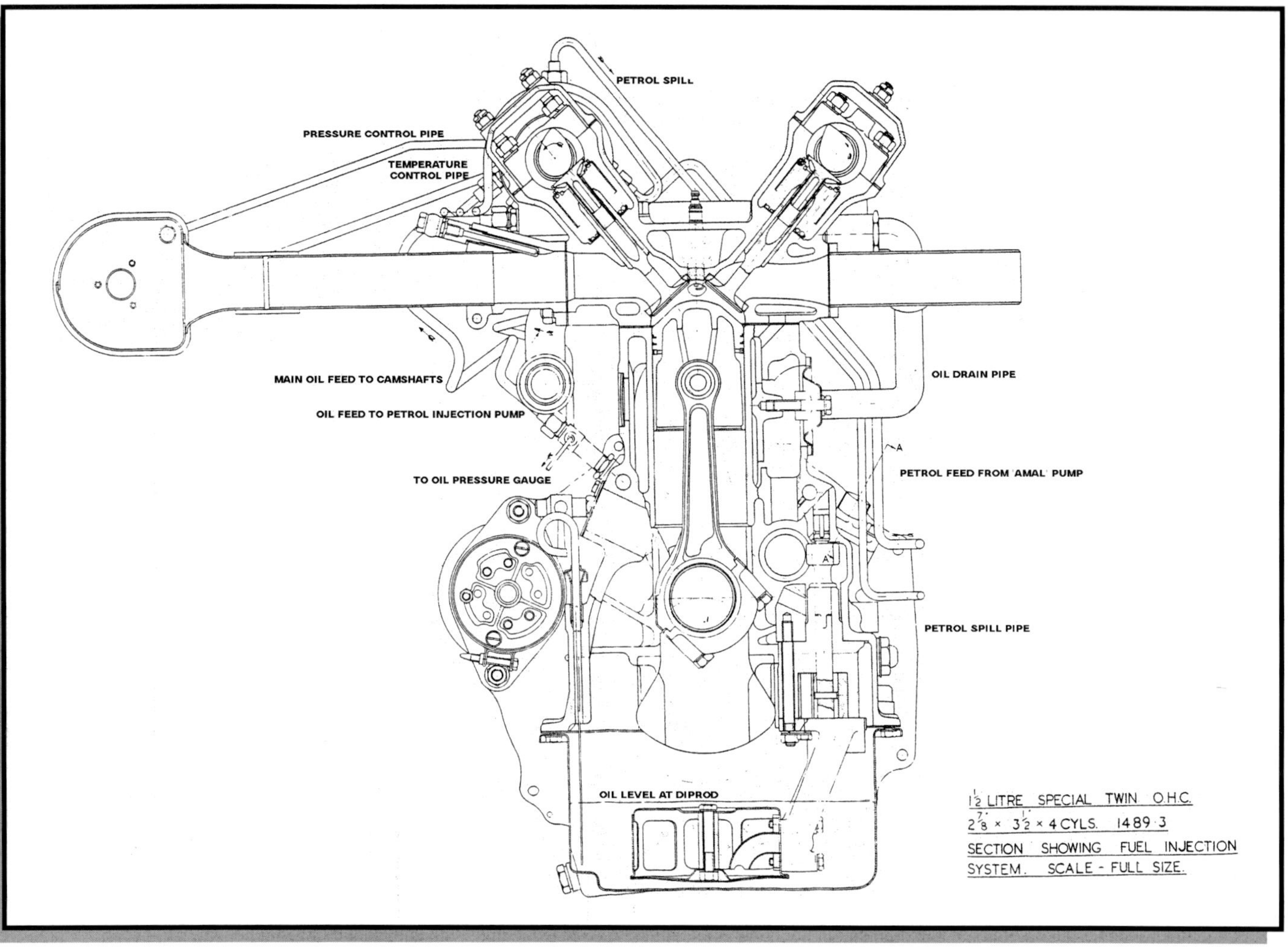

The drawings on these two pages, which have never been published before, show an attempt to use fuel-injection in the Twin Cam engine, with the pump driven from the back of a camshaft. There are several detail differences to the final engine; for instance the valve stems are threaded internally, (Hispano-Suiza style) to take the actuating discs screwed into them. This scheme was abandoned because the base of the shank was prone to cracking. (Courtesy Gerald Palmer)

valve heads opposed at an included angle of 90 degrees, like the Alfa Romeo engine." said Palmer. Alfa had been fitting twin cam engines as standard since 1929 and Palmer was a great admirer of them. "I'm not too proud to admit that I had a jolly good look at the Alfa engine." he said. He spent considerable time at Brooklands on private work at the workshops of Thomson and Taylor, the Alfa agents, and recalled borrowing valves and other engine parts to study when he was drawing up his new engine.

The original 90 degree angle was reduced to 80 in development, again - he thinks - because of intervention by Harry Weslake at Longbridge who maintained that there was a better gas flow if the included angle was smaller. The valves in the Appleby engine were opposed at 66 degrees, which made it a taller unit. Carburation was by twin choke Webers.

In the new 'rationalised' BMC there was no point in producing two engines for a sports car, so an opportunity was sought to pit them against each other. It came a week before the opening of the 1955 London Motor Show, where the MGA was to make its first British appearance.

Three cars were entered for the Ulster TT at Dundrod on September 18, two of them Le Mans cars and the third either the spare Le Mans car or a production car built to the same standards. The racing department at Abingdon installed the twin cams in two of the cars - the Austin engine was put in the car which had finished 17th at Le Mans. It was slightly taller than the Morris engine, so a bonnet bulge had to be incorporated. The Morris engine went into LBL 301, the car which had finished 12th. Alterations were made to the panelwork at the front of the car to try to make it more

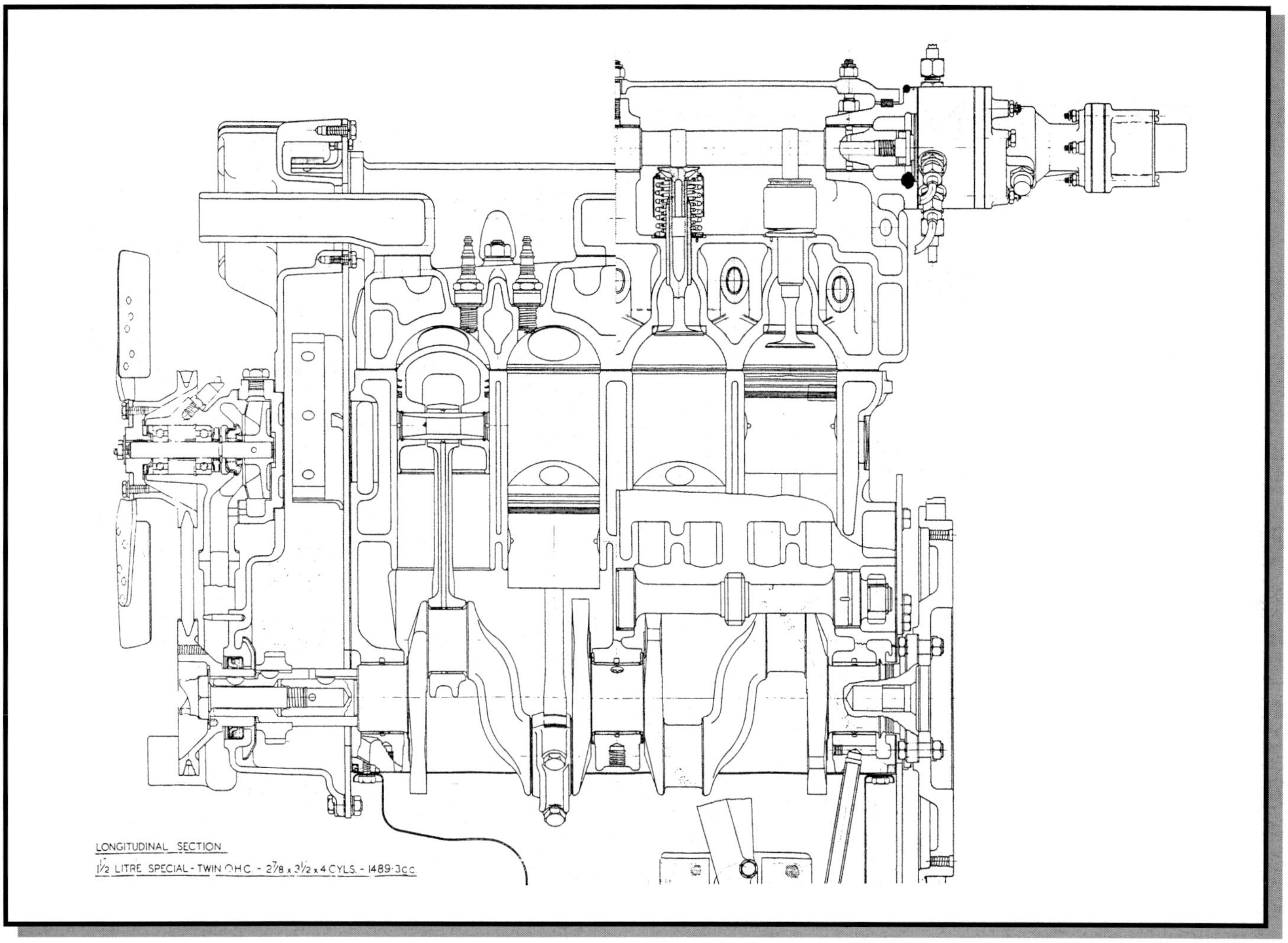

Other differences included the cubic capacity, shape of the piston crown and position of the dipstick.
(Courtesy Gerald Palmer)

aerodynamic, giving it a slightly droopy look at the front.

Just before the cars were about to be shipped to Northern Ireland, the Austin engine was removed and a B-series pushrod version substituted. Some mystery surrounds this last-minute decision, apparently taken by Syd Enever because the Austin engine did not deliver enough power, possibly because the 66 degree valve layout restricted the valve size. It was said to deliver no more power than the tuned B-series unit. Abingdon legend has it that the MG racing mechanic, Tommy Wellman, was called out in the middle of the night to change the engines, and the Appleby version - the only one in existence - was crated up, never to be seen again, supposedly ending up in the bowels of the Motor Industry Research (MIRA) establishment near Nuneaton. Could it still be there somewhere, muse some enthusiasts? Hardly likely; there are no cellars at MIRA and the staff say that if it was there they would probably have found it by now.

The other last-minute alteration was to change the carburettors in the Morris twin cam engine from Solex to Weber, which meant fabricating a new inlet manifold. The Webers were substituted because in preliminary tests on the MIRA track the engine had cut out on right-hand bends due to fuel surge.

The Morris-engined car, driven by Ron Flockhart and Johnny Lockett, ran well, lapping at almost 80mph until a misfire halfway through the race forced its retirement after 23 laps. The misfire was originally attributed to plug trouble, but then it was reported officially that the car had developed a hairline crack in the new manifold, which unbalanced the air/fuel mixture and led to piston failure. This might well have been so, though piston failure for completely different reasons was endemic in the engine's later development.

Of the other two MGAs, the Ted Lund/Dickie Stoop car, LBL 303, retired on lap eight with a split in its

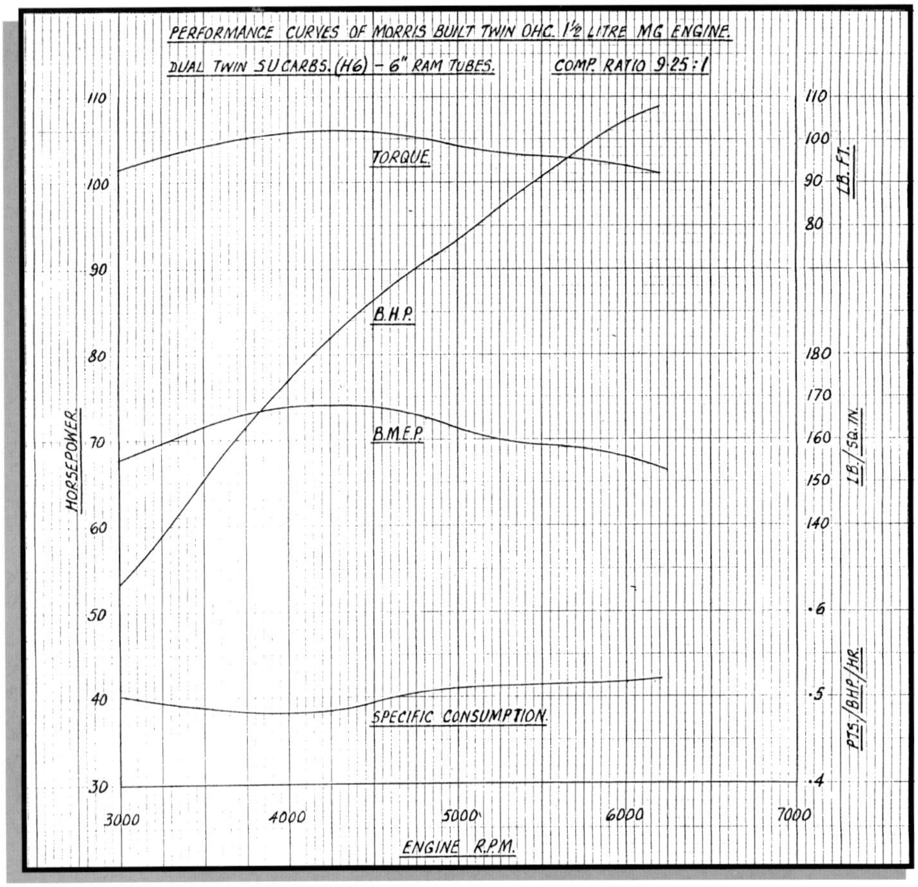

A dynamometer test of the 1489cc engine fitted with experimental twin-choke SU carburettors showed figures similar to the larger production engine. (Courtesy Gerald Palmer)

Gerald Palmer, then 86, at his home in Oxford looking over some of the Twin Cam drawings.

The Dundrod TT Twin Cam with specially-adapted front wings to give better aerodynamics. The front lights would not have been legal on the road as they were too low. (Courtesy BMIHT/Rover Group)

experimental alloy fuel tank. The other - from which the Austin engine had been removed - finished a creditable fourth behind three Porsches.

It was to be another three years before Thornley's vision of a production competition MGA with the twin cam engine was fulfilled. During those years he persisted with his battle with the BMC hierarchy to build it, often against a great deal of opposition. The thrust behind his argument was that the project was not just to make a quicker car, but to compete with Porsche (regarded by MG in the fifties as a threat on both road and track) in MG's major market, the United States. As early as March 1953 Thornley was telling the board: "From the point of view of a competition car in the 1.5-litre class, the car we have to beat is the Porsche 1.5-litre. This has an engine of 1488cc with 80mm bore and 74mm

In the pits at Dundrod, the Twin Cam and one of the pushrod MGAs, which were pitted against each other; both retired before the finish.

stroke which, with a compression ratio of 8.2 to 1, delivers 70hp [sic] at 5000rpm. With a back axle ratio of 3.56 to 1 and a tyre size of 500x16, the makers claim 110mph".

At the time he wrote this Abingdon was still producing the TD which gave out 54bhp and had a top speed of 80mph. The difference between the two cars was enormous and the TF which followed hardly narrowed the gap.

After getting the MGA into production, Thornley kept hammering away at BMC. He wrote in May 1956: "We can materially boost the sales of the ordinary 'cooking' MGA, if we produce a high-performance sports car to go with it". Thus, the Twin Cam was, amongst other things, a marketing ploy

Thornley didn't want to build a car like the Porsche - he just wanted its customers. "I am very content for some time yet with the MGA chassis as a basic design", he wrote to the board in May 1956. "The only thing I covet about the Porsche is its market".

Using one body - the famous 356 and 356A, with three alternative tops - and three different engines, Porsche was able to market eleven different cars. The 1600S, the most direct competitor to the MGA in 1956, retailed in the US for $3470 (£980), a thousand dollars more than the Abingdon car; the top of the range 1500GS four cam coupé retailed there at $4790 (£1350). Even at these prices, Porsche was selling 60 a month in the winter in California. "We have them stone-cold on price", wrote Thornley, " I know from personal conversation that the announcement of the MGA in Frankfurt had them really frightened. We must follow up on this advantage".

The west coast of America was, of course, MG's premier market, and the competition was fierce. Porsche were overpriced compared not only with MG but also Jaguar and had been pushed by their US distributor, Max Hoffman, into producing a 'cheap' 356, the Speedster, to compete not only with MG, but also the TR2 and Austin-Healey. The stripped-down Speedster became a cult car associated with James Dean and made a great name for itself on the race tracks of the west coast. But profits on the car were so small that it was dropped by 1958.

The reputation, however, of Porsche as a builder of fast, well-built sports cars was made. The company traded heavily on the German reputation for quality - offering an enamel badge to owners who did 60,000 miles without a bottom end overhaul and a gold watch to those who did 100,000.

Abingdon was hardly in the business of handing out gold watches and Thornley's US idea was to plan a two-engine, three-body strategy to compete with Porsche. This would comprise the twin cam and pushrod engines with body options of the roadster, the coupé (which was then being prepared for its debut at the 1956 London Motor Show), and a cabriolet top which Vanden Plas was asked to develop, though this was never seen in public.

The Twin Cam coupé could be sold in the US for $2995 (£868), considerably undercutting the expensive Porsches. As for performance, the Porsche 1600S, despite having a 75bhp engine, struggled to reach 100mph, whereas the Twin Cam would be marketed as a 120mph sports car.

The economics at Abingdon were that the MGA cost BMC £460 to produce, £70 of which was the cost of the pushrod engine. It was estimated that the twin cam engine was twice as expensive to produce at £143; other modifications like disc brakes, Dunlop Road Speed tyres and a new bonnet top (necessary to clear the cam boxes) would add another £50. At this stage, it was envisaged that twin cam engine production would run at 50 units a week and Thornley suggested it be installed not just in the MGA but also in a GT version of the Magnette, so engine production would be split evenly between the two. (There is some evidence that he had wanted to keep production as low as 25 units a week, but BMC forced the pace to achieve economies of scale and production eventually went up to 75 a week).

The prototype engines were quieter and sweeter-running than the highly-tuned pushrod in the MGA, it was reported in May 1956. However, even after two years of experiments, there was still a lot of development work to be done on the engine before it was suitable for production. The joint facings were not good and oil leaks were legion, particularly once any part of the engine had been disturbed.

The development work went on and the engines were tested at very high speeds on the salt flats at Utah in MG record-breakers EX179 and EX 181, whose story is told later.

The engine's designer, Gerald Palmer, had long since departed for Vauxhall. At Morris Engines in Gosford Street, Coventry - the old Hotchkiss works where the first-ever MG engine was produced - Eddie Maher and Jimmy Thomson were working on the production version, for BMC's chairman, Len Lord, had finally agreed that it could go ahead.

The Austin engine was forgotten; not only had it not performed well, but the cost of putting it into production was far greater since it employed a brand-new bottom end and block.

Lord stipulated that whilst the Morris engine could have the new cylinder head, the B-series block would have to be used, as in Palmer's design. This meant the block would be machined on the Transfermatic equipment on the Longbridge line which was producing the unit for the rest of the empire. It was a half measure, for it tied MG to the B-series bore centres, and was not the economy it seemed, for while the other BMC 1.5-litre cars - including the pushrod MGA - were using the 1489cc unit, the Twin Cam engine had grown to 1588cc to take advantage of international competition regulations, specifically Appendix J of the International Sporting Code, which now allowed for a class of up to 1600cc. To compete in the American market it was essential that MG went close to the limit - the Porsche threat was potent.

The decision meant that while the stroke was the familiar 88.9mm, the bore was increased from 73mm to 75.4mm, so it still had to have additional machining on a special line at Longbridge before being despatched to Coventry for the rest of the build-up, which began at the rate of five engines a day.

MG was to have its first overhead cam engine since the 1936 demise of the single overhead cam unit in the PB Midget. Expectations were high ...

"Surely, it's not beyond the wit of man to cure piston trouble" - Gerald Palmer

Engine

On Gerald Palmer's drawing board at Cowley the twin cam had started life as a unit of 1489.3cc, the same bore and stroke dimensions as the existing B-series engine, but, as explained elsewhere, by the time it was in production, the dimensions had grown to 1588cc.

The bore was enlarged from 73.02mm to 75.4mm, an increase achieved not without sacrifice, since the bores became 'siamesed' - that is the water space disappeared between cylinders 1 and 2 and cylinders 3 and 4. The new head - still a thing of beauty today - was of aluminium alloy with the MG octagon on each cam cover picked out in red and the sparkplugs spaced down the middle. The camshafts ran in three white metal bearings, with the cams pushing down on upturned bucket tappets made of Brimol cast iron, which were returned by double valve springs. These, in turn, opened the large-diameter valves - the inlets with a head diameter of 40.38mm and the exhausts with 36.58mm. The exhaust valves were made of K.E. 965 steel with stellited faces and the valve seats were cast austenitic iron, machined in the head.

The combustion chambers were hemispherical, of the Riley type (not surprising since the engine had come from the MG/Riley design office). The spark plugs running down the centre of the head were offset to allow the maximum space for the large valves.

The drive to the camshafts - inlet shaft on the right of the engine, exhaust on the left - was ingenious. It began at the crankshaft nose, behind the dynamo pulley with a case-hardened helical gear - straight-cut gears were used in the prototype engines, but helical were adopted for production to reduce noise. The gear drove a jackshaft located where the camshaft would have been in the pushrod ver-

A Twin Cam engine removed for overhaul. Note the position of the distributor on the bottom right, seen here without the cap.

sion of the B-series engine, driving it at half engine speed. The jackshaft carried a sprocket which drove a long double-row chain anti-clockwise up and across the 25-tooth camshaft sprockets, directed by an idler sprocket on the drive side and an adjustable sprocket on the other. This shaft drove not only the timing chain but also the distributor at its front, oil pump at its centre and rev counter at the rear.

John Thornley's vision of a fuel injected engine was dropped in the development stage, probably because of cost and reliability, since fuel injection in the fifties was a much less certain art than it is today. It did not begin to appear on production cars in Britain for another ten years.

But at least one injected engine was planned, specified with an Amal

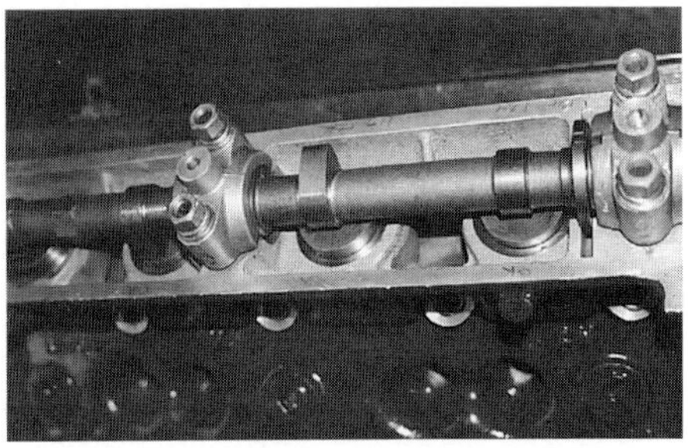

The way the cams operate on the upturned bucket tappets can be clearly seen.

fuel pump and an SU injection unit driven from the rear of the inlet camshaft. In the end, however, twin SU HD6 $1^{3/4}$ins diameter semi-downdraught carburettors, mounted at an angle of 22.5 degrees, were fitted.

The camshafts run in three white metal bearings.

The float chambers, fed by an SU high-pressure pump, were mounted flexibly to reduce vibration and the rear throttle had a slight lead over the front to provide a more progressive take-up of power.

It's worth noting that when Palmer tested a prototype engine of 1489cc using experimental twin choke SU carburettors with 6 inch ram pipes, he achieved 109bhp at 6200rpm, compared with the production car's 108bhp at 6700rpm with the bigger 1588cc engine.

The inlet manifold gave smooth porting to the mixture, and on the other side of the engine each cylinder had its own exhaust pipe leading to a union of cylinders numbers one and four into one pipe, then two and three into another.

The original pistons had raised domes, resembling cones with their tips cut off, which were necessary to achieve the very high compression ratio of 9.9:1, and the connecting rods were much stiffer than on the standard B-series. The forged three-bearing crankshaft was made of EN 22 steel, heat-treated to 85 tons tensile strength.

The mains and crankpins were the same diameter as on the B-series, but the mains were narrower to allow for thicker crank cheeks. Heavy-duty lead indium bearings were used.

The sump, of finned aluminium alloy, had a capacity of 13 pints - double that of the MGA 1500, and evidence that the car was expected to

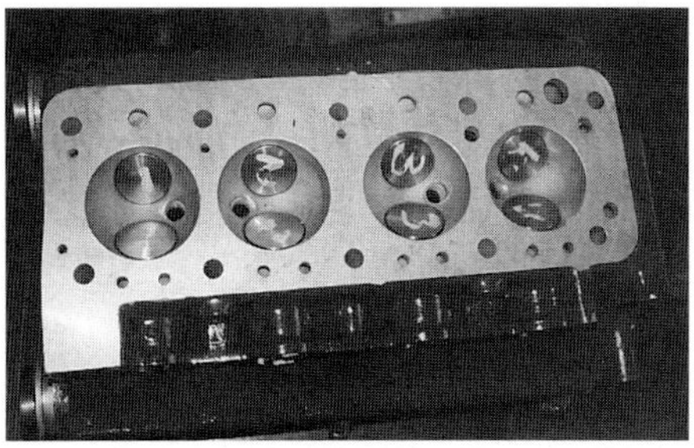

Large diameter valves gave good breathing.

be used in some anger, though if this were not enough an oil cooler was offered as an optional extra. "Particularly suitable for hot countries" remarked the BMC briefing, unsurprisingly.

The engine, being longer than the

Dunlop rear disc brakes were an innovation for MG. The rear suspension - leaf spring and lever arm damper - carried through to the end of MGB production in 1980.

pushrod version, also forced a change in the cooling arrangements. As the ordinary radiator with header tank would not fit under the bonnet, a shorter radiator was used to which water was ducted from a cast aluminium header tank located to the side of the engine above the exhaust manifolds. The extra weight, compared with pushrod engine, amounted to about 50lbs (22.7 kilos), but it delivered 58 per cent more power and the engine speed at which it peaked was 6700rpm rather than 5500rpm.

Transmission

The existing transmission was well able to handle the extra power. The B-series gearbox - ratios unchanged - was installed, but close ratio gears were offered as an option from launch. The rear axle ratio of 4.3:1 was also carried over, though 4.55:1 was available at an extra £10 2s 6d.

Below, The knock-on wheels were heavily dished to allow clearance for the disc brake calipers behind.

Chassis

Certain chassis details were different and dealers were warned by BMC that it was impossible to convert the standard MGA to Twin Cam specification, though several people have subsequently done so.

The main differences were that because of the longer engine, the rack and pinion steering was moved an inch (25mm) further forward and utilised longer steering arms. The front track was slightly wider - about half an inch - and the front coil springs were stronger to cope with the heavier engine.

Brakes

Much was made of the Twin Cam's $10^{3/4}$in diameter four-wheel Dunlop disc brakes "famous for their success in racing", as the publicity put it. They were also used on a handful of sporting cars such as Jaguar and Alvis.

MG dealers were told that disc brakes fitted to the Twin Cam would *not* be made for the basic MGA. This turned out to be only partially true, since, less than a year later, 11ins Lockheed discs were fitted to the front of the 1600. These turned out to be a great deal more successful than the Dunlop version, since there was no attempt to use discs on the rear wheels. A besetting problem of these brakes was getting the handbrake adjustment right for the rear pads to bind on adequately the discs. There was, surprisingly, no servo assistance.

Wheels and hubs

One of the distinguishing features of the car was its Dunlop centre-lock wheels, similar to those used on the D-type Jaguar and BRM. Many writers have said that they were chosen for aesthetic reasons, and because they were lighter than wires, but the real reason for the heavily-dished disc shape was to enable the wheel to clear the bulky disc brake calipers, a fact explained to dealers in their briefing papers before launch. The wheels were located on the hubs by four dowels, or pegs, rather than by splines. Front hubs had Timken taper-roller bearings instead of ball-races.

The body

The Twin Cam was offered as both roadster and coupé from launch. Little distinguished the Twin Cam bodies from the standard product, other than

The Twin Cam prototype photographed in the factory. The badges have yet to be fitted behind the air vents. This car was sold at auction in 1995 for well over £20,000.

Below: A wolf in sheep's clothing; the distinctive wheels and badging were the only external features that distinguished the Twin Cam from the pushrod car.

Twin Cam motifs beside the bonnet ventilators and beneath the octagon on the boot. There was a subtle change to the bonnet panel to accommodate the slightly taller engine. The bonnet length was increased from 47$^{7/8}$ins to 48$^{3/8}$ins to allow for greater curvature above the engine. This length bonnet was also fitted to the pushrod cars.

Apart from other very minor changes, the most noticeable feature inside the roadster was the vinyl-covered dash panel which matched the trim, already specified in the 1500 coupè, and later adopted in the 1600 Mk II roadster. The rev counter read up to 7500rpm. The radio speaker grille surround was chromed and there was a chrome strip along the bottom edge of the fascia.

So-called 'de luxe' seats were offered as an extra, and were wider than the standard offering.

One of the more unusual extras was a competition windscreen made in shatter-proof plastic. Of the same width as the standard screen, it was only 6$^{1/2}$ins deep, substantially reducing drag at high speeds.

In production

"Splendid fun, but noisy and an oil drinker" - Motor Sport

Despite a very long lead time, problems with the Twin Cam engine were not resolved until it was almost out of production. BMC Service Bulletins on piston problems were still being issued when the last car had left the line! As Bill Boddy commented wryly in *Motor Sport*: "Abingdon made rather heavy weather of adapting Henry's 1912 type of overhead valve actuation to the

MGA".

In fact, there can be few cars released to customers with such a likelihood of known and expensive failure. Many excuses were given: it was a competition car driven by the wrong sort of people; "Vicars were doing their shopping in it", one ex-Abingdon employee told me; "It was not maintained properly", said another, "particularly in America". The truth was that, despite being under constant development from March 1953 onwards, the engine was still not ready to be put on sale five years later in the summer of 1958.

Not that this reflected in the MG briefing material sent to dealers in July 1958, which said of engine development ... "it is only now, when the MG engineers are fully satisfied with the engine, that it is being offered to the public".

Prototype testing had first of all shown up overheating problems caused by steam pockets in the head. These were cured by running a pipe into the centre core plug from the header tank. Then the piston problems appeared. On runs at the MIRA proving ground, one tester remembered undoing a sparkplug and seeing the sun shine on the top of the con rod because the piston crown had disappeared. The Champion sparkplug technical team was called in and designed the N4 plug specifically for the Twin Cam, though it was some time before it was fitted; less than satisfactory, it was eventually replaced. The N4 alleviated the piston-holing to some extent, though the N3 was recommended for high-speed work, and this became standard in May 1959.

This 8-page booklet told owners - in great detail - how NOT to blow up their car's engine!

The piston burning was largely caused because the very high compression ratio of 9.9:1 made the engine very vulnerable if low octane fuel was used or the ignition was too far advanced.

Accompanying this weakness was very high oil consumption - as much as a pint every 120 miles. The pistons ran in chrome molybdenum bores with clearances of 9 thou, which also made the engine noisy.

Despite these unresolved difficulties, the first car, chassis number 501, came off the body line on April 22 1958. It was finished in green, with black trim and a blue hood. Like 502 which followed, in red body and trim, it was a factory demonstrator. There were then three cars produced for the home market and 15 for North America.

Road & Track in the US had wanted to test a Twin Cam since seeing the engine in development in 1955. But, after seeing the cars in July, they put off the test, reporting: "The first few samples were rushed over and they just weren't quite right as delivered." Ominous words ...

By the time the car was due for launch on July 14 1958, 35 units had been built. *Autocar* and *Motor* had been given a chance to carry out extended tests on the Continent beforehand, one in Belgium and the other in Germany. Other, less important, publications were given an opportunity to hammer four pre-production versions on the Ministry of Defence proving track for fighting vehicles at Chobham in Surrey. The journalists were given four courses for lunch (it is significant that the lunch menu was headed "The Nuffield Organisation" - this was six years after Nuffield was supposed to have been absorbed into BMC. It was this kind of rivalry with Austin that helped to bring down the empire) and two courses on which to drive - the high-speed and the 'snake'. After they hurtled around, MG fitters anxiously checked oil levels after every few laps.

The BMC press briefing proclaimed proudly: "The MGA Twin Cam is a great British challenge to the world of motor sport and, with its speed of two miles a minute and disc brakes all round, will earn MG a brand new name in keeping with its old tradition".

Autocar thought much the same, saying that the price required for the extra performance and increased braking was not unreasonable, and adding: "This latest product from Abingdon must rank among the world's outstanding sports cars". *Autocar's* testers, taking advantage of the free-revving engine on their Continental journey, repeatedly took it up to the recommended maximum of 7000rpm; they had been warned that valve crash set in at

The first and last Twin Cam

The last scheduled Twin Cam, number 2610, came off the line at Abingdon on April 13 1960. But six weeks later, Mike Ellman-Brown persuaded Abingdon to build him one final car.

Ellman-Brown had been taking part in competitions in a pushrod MGA and was waiting for the bugs in the Twin Cam to be sorted out before buying one. When he heard that production was being stopped he contacted MG's general manager, John Thornley, who agreed to put one final Twin Cam down the line which was now making only pushrod cars.

Factory records show that 2611 was built to Thornley's order and finished in a non-standard colour - Woodland Green. It was fitted with the close-ratio gearbox and a 4.55:1 rear axle. Another unique feature was that it was photographed throughout its journey down the line (pictured in *Making MGs*, Veloce 1995), which ended on June 14 1960, when Mike Ellman-Brown went to the factory to pick it up - something owners were allowed to do in the fifties.

In 1996, the car was still in his garage in Kent, having done some 9000 miles, but with frontal damage sustained when the handbrake slipped and sent the car on an unscheduled journey. The Dunlop pads, never of much holding power on the rear discs, lost their grip when the discs cooled and contracted. It is to be restored.

By coincidence, some 15 miles away resided the first Twin Cam, the prototype ORX 885, first registered in January 1958 and owned by the factory for four years. When Don Smith bought it at auction in 1995 to join his other Twin Cam, it had been extensively restored. The body is all aluminium and the engine, breathing through twin Weber carburettors, develops 160bhp. It has been extensively campaigned in club racing and, at the time of writing, has a Quaife limited-slip differential.

7600rpm. They noticed engine vibration at 2500 and 5500rpm, but pointed out that 80 to 90mph could be held as a cruising speed and that the car was quite happy at 100mph for long stretches.

But they also noted that the engine was noisy and neither smooth nor silky and used "a considerable quantity of oil" - five pints during one 800 mile journey. The distributor became covered in oil during the test and the dipstick was difficult to reach. Roadholding and cornering were praised, as were the Dunlop disc brakes, and there were minor criticisms of the heat build-up in the footwells and headlamp range.

In summary, however, *Autocar* concluded that the Twin Cam maintained the MG tradition of good looks coupled with a very fine performance.

Motor's verdict was rather similar and it described the car as being "... relatively noisy, entirely tractable and will probably appeal in the main to owners who wish to win races". It noted that of the cars tested, only those built specifically for sports car racing could keep pace with the Twin Cam in standing starts to speeds of 60, 70 or 80mph.

The Twin Cam was fun to drive, had a first-class gearbox (though closer ratios would be desirable) and the Dunlop disc brakes deserved the highest praise.

On the minus side, *Motor* noted the high oil consumption, difficult access to engine ancillaries and the fact that during its 1500 mile test the shock absorbers "virtually disappeared", leading to severe scuttle shake - a fault addressed in production.

The car had been presented from the outset as a limited production competition vehicle. *Road & Track* pointed out that it had been designed primarily to regain prestige for the MG marque in production sports car racing, and was for the genuine enthusiast. But this also meant that the engine, with its high compression ratio, was highly strung, needing careful attention to ignition timing and fuel octane ratings. Two things militated against this - the distributor was very difficult to reach, being concealed beneath the cam cover, even though the points gap needed resetting every 3000 miles, and more than 80 per cent of production was exported to countries where high octane fuel was not readily available.

The distributor accessibility problem was attended to before the first 100 cars were off the line. Louvred detachable panels were fitted inside the wings to give better access to units like the distributor and oil filter. They also improved engine cooling. The design of these panels was the first job given to Don Hayter when he joined MG from Aston Martin. He then went on to design the MGB and became MG's chief engineer.

By November of 1958 the first of many attempts to improve oil consumption was made with adoption of twin-segment scraper piston rings. The following month new pistons and rings were fitted to overcome piston noise, reducing bore clearances from 9 thou to 4 thou.

Because the engines were so free-revving there were occasions when pistons and valves met, with disastrous consequences. This happened particularly when the cast iron bucket tappets fractured or seized in the aluminium head. To cure this two fixes were necessary; the final one was the insertion of an iron sleeve in the head in which the tappet could run.

But the major problem, that of piston burning, was slow to be resolved. In May 1959, Champion N4 or N5 plugs were replaced by N3 to reduce heat at the piston crown, or N58R for hard driving. This did not always sort out the problem, since garages at servicing time were wont to fit N5s to all B-series engines as a matter of course.

Evidence of concern at Abingdon about engine reliability appcared later

The first Twin Cam brochure suggested that the extra performance came cheap, compared with other sports cars, and that the standard MGA could not be converted to the same specification.

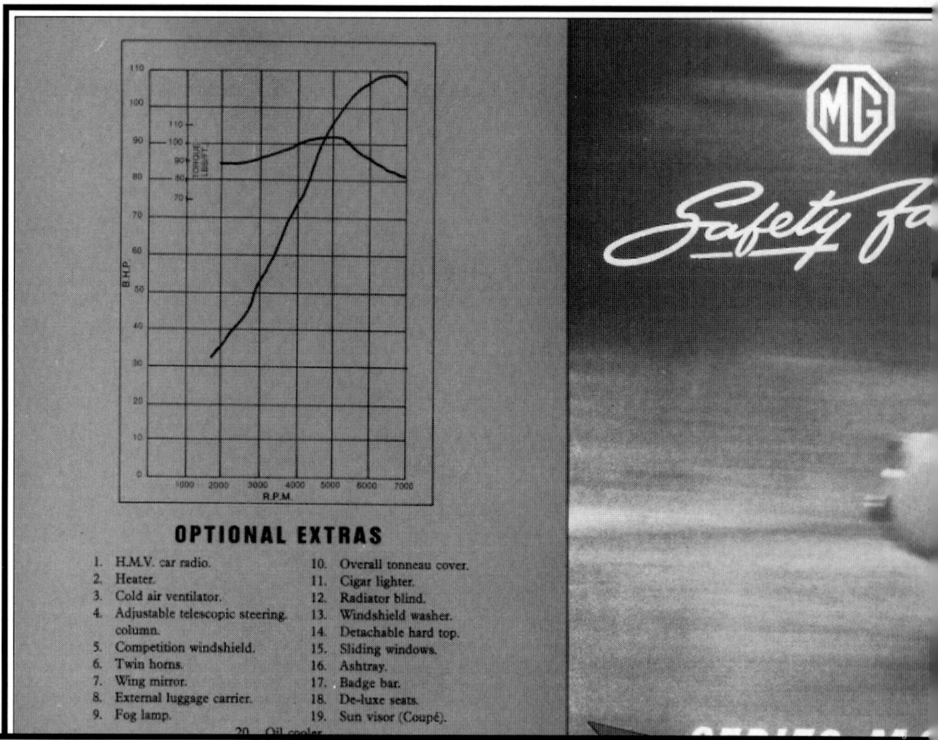

B.H.P.

R.P.M.

OPTIONAL EXTRAS

1. H.M.V. car radio.
2. Heater.
3. Cold air ventilator.
4. Adjustable telescopic steering column.
5. Competition windshield.
6. Twin horns.
7. Wing mirror.
8. External luggage carrier.
9. Fog lamp.
10. Overall tonneau cover.
11. Cigar lighter.
12. Radiator blind.
13. Windshield washer.
14. Detachable hard top.
15. Sliding windows.
16. Ashtray.
17. Badge bar.
18. De-luxe seats.
19. Sun visor (Coupé).

in that year when it was decided that all new owners should be given a booklet entitled *Getting the Best from your Twin Cam* (which might have been more aptly entitled *How not to Wreck Your Engine*) with their service vouchers and warranty packs. The booklet went into some detail about the need for high octane fuel and correct settings:

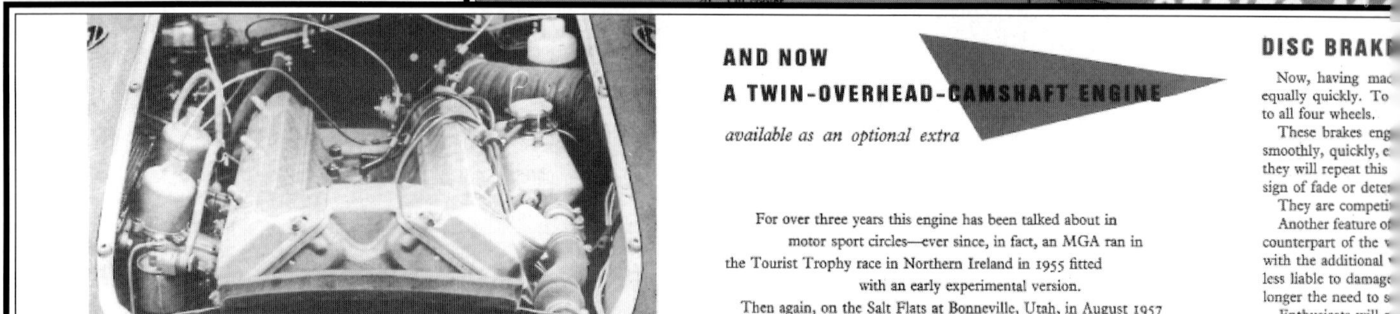

AND NOW
A TWIN-OVERHEAD-CAMSHAFT ENGINE

available as an optional extra

For over three years this engine has been talked about in motor sport circles—ever since, in fact, an MGA ran in the Tourist Trophy race in Northern Ireland in 1955 fitted with an early experimental version.

Then again, on the Salt Flats at Bonneville, Utah, in August 1957 Stirling Moss achieved 246 m.p.h. with the M.G. Special EX181, and this time the twin-cam engine was highly supercharged.

The M.G. engineers have held back the announcement of this long-awaited engine until they are quite satisfied with it.

And now, what do they offer?

POWER AND PERFORMANCE

See the power curve. Note particularly, not only the maximum horse-power, but the substantial flatness of the curve all the way from 5,000 to 7,000 r.p.m.

Note, too, that the engine capacity has been increased to 1588 c.c. to take full advantage of the International Touring Car classification.

The safe maximum engine speed is regarded as 7,000 r.p.m., and with the standard MGA axle ratio of 4·3 : 1 and with Dunlop Road-speed 5·90—15 tyres this should give around 120 m.p.h.

The overall acceleration is very spirited indeed. Using 6,500 r.p.m. as the gear change point, the car can be accelerated from rest to 100 m.p.h. in the order of thirty seconds. Nor is this all. The acceleration time for 0 to 110 m.p.h. is in the order of thirty-eight seconds, which goes to show that, even at 100 m.p.h., the car is still accelerating quite hard.

CONVERSION

To avoid misunderstanding it is here stated that the normal standard M.G. Series MGA with drum brakes and push-rod engine continues unchanged. The engineering changes necessary for the installation of the twin-cam engine and the disc brakes are so extensive that conversion of existing MGA cars is not practicable.

PERFORMANCE OF THIS ORDER USUALLY

very quickly, one must be able to stop it—
e latest-pattern Dunlop disc brakes are fitted

of complete mastery over the car and stop it
convincingly from maximum speed. Moreover,
numerable times in quick succession without a

excellence.
e centre lock disc wheels, which are the modern
se retain the advantages of the knock-on hub
are easier to keep clean and maintain and are
eased efficiency of the disc brake, there is no
ntilation provided by the wire wheel.
se the ideal sports car wheel of the future.

TION

n. (2·969 in.), stroke
c.c. (96·906 cu. in.);
brake-horse-power
t driven by duplex
appets and running
3-bearing counter-
backed lead-indium
with one scraper and
on pins; connecting
l-indium bearings;
feed lubrication by
nt external full-flow
n with thermostatic

semi-downdraught
ers. Rear-mounted
Fuel tank capacity
gallons). Fuel gauge

2-volt oil-filled coil
vacuum and centri-
pment; belt-driven
single-pole positive
arter switch; twin-
win stop-tail lamps
ear reflector equip-
ng headlamps with
ot-operated dipping
s batteries mounted

CHASSIS: Exceptionally sturdy box-section frame, specially braced for torsional rigidity; rear end of chassis swept over rear axle.

TRANSMISSION: Hydraulically operated single dry-plate Borg & Beck clutch, 8 in. (20 cm.) diameter. Four speeds and reverse; synchromesh on second, third, and fourth; overall gear ratios: first 15·652, second 9·520, third 5·908, top 4·3, reverse 20·468. Central remote-control gear change. Tubular propeller shaft with needle-bearing universal joints.

REAR AXLE: Three-quarter-floating rear axle with hypoid final reduction gears; ratio 4·3 : 1.

SUSPENSION: Semi-elliptic rear springs controlled by hydraulic dampers. Independent front suspension by coil springs and wishbone-type links controlled by hydraulic dampers.

STEERING: Direct rack-and-pinion steering with large-diameter spring-spoke clear-view steering-wheel. Left- or right-hand steering according to market (left-hand-drive not available in the United Kingdom).

BRAKES: Dunlop caliper-type disc brakes operated hydraulically on all 4 wheels by a foot-operated master cylinder. The separate parking hand brake mounted on the rear calipers is operated mechanically on the rear discs by a central hand brake lever with press-button ratchet control.

TYRES AND WHEELS: Dunlop 5·00—15 Roadspeed tyres with tubes. The disc wheels are centre-lock type ('knock-on').

INSTRUMENTS: Large speedometer with dead-beat reading and headlamp high-beam warning lamp; large revolution indicator with ignition warning light; oil pressure gauge; water temperature gauge; fuel indicator gauge; ignition switch; rheostat panel light switch; mixture control; map-reading light; direction indicator switch and warning light; lighting switch.

BODY DETAILS: Open 2-seater streamlined body with enclosed luggage boot; adjustable bucket-type seats, leather upholstery with leathercloth on non-wearing parts; door pockets; safety-glass windshield; folding waterproof hood with large rear transparent panels; 2 detachable sidescreens with combined stowage and hood cover; driving mirror centrally situated; spare wheel, tools, jack, and starting handle housed in rear boot; quick-release petrol filler cap; remote-control locks for bonnet and luggage boot lid; one-piece bonnet hinged at rear, giving easy access to engine unit.

COLOURS (2-seater): Black with Red or Green upholstery; hood material Ice Blue or Black. Orient Red with Red or Black upholstery; hood material Black. Ash Green with Grey or Black upholstery; hood material Ice Blue. Glacier Blue with Grey or Black upholstery; hood material Ice Blue. Old English White with Red or Black upholstery; hood material Black.

COUPÉ: Black with Red or Green upholstery. Orient Red with Red or Black upholstery. Old English White with Red or Black upholstery. Ash Green with Grey or Black upholstery. Mineral Blue with Grey or Black upholstery.

TWICE THE MONEY!

"Particular care is needed with this engine owing to its high compression ratio (9.9:1) which makes it extremely sensitive to variations in fuel, ignition timing and the heat range of the sparking plugs ... variations in these settings will increase the combustion temperature and if the variation is excessive, pre-ignition will cause high shock waves resulting in damage to the engine."

And the last sentence in heavy type exclaimed: **"Under no circumstances should the ignition be advanced beyond TDC."** (Top dead centre).

All this was unlikely to endear Twin Cams to their new owners, with the factory putting the onus on them to keep their engines in one piece. Belatedly, and in short order, in early 1960, a new distributor was introduced, without a vacuum advance, low compression pistons of 8.3:1 were fitted, spe-

Twin Cam coupé deliveries

		1958	1959	1960	TOTAL
Right-hand drive:	*home*	26	115	9	150
	export	7	15	2	24
Right-hand drive: (South Africa)	CKD	0	24	0	24*
Left-hand drive:	*export*	5	24	1	30
Left-hand drive:	*N. America*	10	85	0	95
TOTAL		48	263	12	323

*These figures are courtesy of the BMIHT. *South African build figures put this at 28.*

cifically for those areas where premium grade fuel was not available and the octane rating was below 95. This piston change reduced power by 8bhp, but it solved the original problem and the Twin Cam became as reliable as any other sports car.

As Bill Boddy of *Motor Sport* wrote after a Goodwood TT meeting in the summer of 1960: "... these Twin Cam MGs have now achieved a satisfactory degree of reliability".

It was, however, too late. The car had become notorious, particularly in America, for being unreliable. Marcus Chambers, BMC's competitions manager, reported that it was doing damage to MG's reputation in the United States. BMC's bosses, who had extolled the Twin Cam's virtues in the company's annual report for 1958 as "a car of outstanding performance for competition work", put a stop to it in May 1960.

But, even before then, there had been some back-pedalling. In December 1959, John Thornley was writing in the MG Car Club magazine *Safety Fast!* that the Twin Cam had set out to be "a darned good Grand Touring car", rather than one which could live with expensive lightweight machinery (for example, the Porsche).

It all ended badly and sadly. John Digby, who worked at the time for BMC's press office, remembers seeing the unwanted twin cam heads being smashed with a sledge-hammer at the back of the scrap stores at Abingdon. What wouldn't one give for a new original head now?

There is some truth in the view that the wrong sort of people bought Twin Cams. Some bought them for their wives as shopping cars and they proved intractable when used regularly in city traffic. One owner appealed in *Autocar* in May 1961 for help to solve his plug-fouling problems. He had a lengthy correspondence with the factory who had offered various suggestions, none of which had worked. His local MG dealer in Kingston, Surrey, refused to adjust the carburettors, referring him to the London distributor. Two letters to them had gone unanswered ...

Peter Wood, who has specialised in restoring Twin Cams for more than 30 years, said the cars were often bought as 21st birthday presents. "You'd get lads revving them to 7000 and beyond, blowing up the engine then having it replaced under the warranty, which BMC always respected.

"I knew one lad who went through three Twin Cam engines in as many months. They should have done what Lotus did and put a rev limiter on the distributor".

The replacement engines were often described as Gold Seal reconditioned replacements, but they were always new rather than remade.

Perhaps it is the case that too many of these highly-tuned cars fell into the wrong hands and were treated unsympathetically. It had been John Thornley's view that only 25 a week should be produced, for a specialist market, but there is some evidence that BMC found such low production runs inefficient and insisted on tripling the number of cars built.

It was an unhappy experiment which ended in disappointment and damage to MG's reputation. Wilson McComb, the MG historian, described it as a "much-maligned car". Yet, paradoxically, the surviving Twin Cams are the most valued of Abingdon's post-war products. The prototype, ORX 885, was sold at auction in London in 1995 for well over £20,000.

Part of the reason for their higher value is, of course, their scarcity, as only 2111 were made over two years, of which some 360 were sold in Britain. Many had the engine replaced by a pushrod version under warranty and others were replaced when parts became difficult to get. So now, in the UK and USA together, there are probably around just 100 roadworthy cars left.

The real value of the Twin Cam lies in its charm; it is a glorious car to drive and the beautiful red-painted engine, with its polished alloy cam covers bearing the MG octagon picked out in red, is a constant delight to those few lucky enough to own one.

Twin Cam - performance comparisons

These comparative figures were handed out to the motoring press at the car's launch. The Twin Cam figures were based on what was called the "MGA Road Test Department - Prototype Test Results".

The 'Basic MGA' figures, as they were described, were taken from the *Motor* road test of the coupé in August 1957. The coupé was a faster car than the standard roadster - possibly up to 5mph quicker at maximum speed - and this particular test was done at the Montlhéry circuit, which was hardly typical of road use.

	Twin Cam	**Basic MGA**
Capacity	1589cc	1489cc
Bore & stroke	2.96in x 3.5in	2.87in x 3.5in
Compression ratio	9.9:1	8.3:1
Maximum bhp	108 @ 6700	72 @ 5500
Maximum torque	105 lbs.ft @ 4500rpm	80 lbs.ft @ 3800rpm
B.M.E.P	163.lbs.ft* @ 4500 rpm	133 lbs.ft* @ 3800rpm
Maximum rpm	7600	6000
Maximum speed	120 mph	103.8 mph
Acceleration through gears		
0-30mph	3.75 secs	5.0 secs
0-50mph	8.95 secs	10.8 secs
0-70mph	14.55 secs	21.4 secs
0-80mph	18.00 secs	32.1 secs
0-98mph [sic]	24.55 secs	-
0-100mph	30.4 secs	-
0-110mph	37.7 secs	-
Maximum speed in gears:		
1st	33mph	26mph
2nd	54mph	42mph
3rd	86mph	68mph
Overall fuel consumption:	23-28mpg	27-31mpg

**As printed*

When *Motor* tested the Twin Cam in Germany before its announcement they achieved figures very different to the factory's.

	MG factory figures	**Motor road test figures**
Acceleration through gears:		
0-30mph	3.75 secs	2.6 secs
0-50mph	8.95 secs	7.3 secs
0-70mph	14.55 secs	12.3 secs
0-80mph	18.0 secs	16.2 secs
0-98 mph [sic]	24.55 secs	*(0 - 90)* 24.6 secs
0-100mph	30.4 secs	40.3 secs
0-110mph	37.7 secs	-
Maximum speed in gears:		
1st	33mph	31mph
2nd	54mph	50mph
3rd	86mph	81mph

continued overleaf

Maximum speed:	120mph	115mph
Overall fuel consumption:	23-28mpg	22.2mpg

Autocar figures were different again, apart from the top speed, which agreed with those of *Motor*, but they timed the 0-70mph at 17.3 secs and 0-90mph at 30.0 secs.

Road & Track's figures almost paralleled those of *Motor*. They ignored the recommended 6500rpm as the best gearchange point, to which *Motor* had adhered, and went for changes at 7000rpm, the stated maximum.

Which all goes to show that road testing with pre-production cars in those days was an inexact science, though the uncharitable, looking at MG's own figures, might think that they were striving rather too hard to meet the magic two-miles-a-minute.

The de Luxe

The untimely end of Twin Cam production left the Abingdon factory with an overstock of the parts that made the car distinctive - the engine, Dunlop disc brakes and wheels. Many of the engines were witnessed being broken up and the remaining brakes, wheels and chassis were used in an uncatalogued model that had the standard MGA 1588cc pushrod engine and was known unofficially as the de Luxe: it was known even more unofficially as the Competition MGA and sometimes even the MGA 1600 Mk IIb.

Sales brochures published after the Twin Cam had gone out of production offered these brakes and wheels as optional extras to the pushrod model. The Twin Cam chassis, which although superficially the same as the MGA 1600, had among other differences an alteration in brake and clutch pedals. Altogether there were some 190 parts changes compared with the standard model. Several owners who opted for the disc brakes and Dunlop wheels on a MGA 1600 ended up, probably unknowingly, with a Twin Cam chassis.

Using up the left-overs, the first car, 91240, a works demonstrator, was finished on June 9 1960 and, between the end of June and July of that year, another 21 were completed, 12 of them for the home market.

Coupé and roadster versions were built over the next two years, spanning the change from the 1600 to the 1600 Mark II. The first de Luxe cars using up the Twin Cam bodyshells had the removable panels in the front wheelarches which allowed access to the engine. Once the Twin Cam shells were used up, however, the model reverted to a standard bodyshell.

Some cars were built to quite high specification. The factory records show, for instance, that a Mr J W Keller had a white coupé, car 102042, built in June 1961 with Road Speed tyres, radio, oil cooler, anti-roll bar and badge bar.

Production continued in small batches of 20 or so - a typical example was a run of 19 for the North American market in July 1961 from car number 102557 to 102575. The disc brake option there cost $500 extra. 63 per cent of all de Luxe production crossed the Atlantic. The last run was a batch of some 70 roadsters to North American specification - some with close ratio gearboxes - which were completed between May and June 1962, ending with car 108652.

One of these US cars, with close ratio gearbox and high compression head, originally sold by Morris Garage of Oneota, New York State, was resold in 1994 with only 13,011 miles recorded. It still had its original Michelin X 165/15 tyres.

De Luxe production ended the same month as the MGA finally went out of production and in total 395 were built in the UK - according to Anders Clausager, archivist of the British Motor Industry Heritage Trust. Just over 80 were the MGA 1600 version and the rest the Mark II version. Thirty-five of the total number built were coupés, which makes them the rarest of all MGA models.

There's been speculation that this total production figure of 395 was probably governed by the original order for Twin Cam chassis parts, which could have been around 2500 sets. Thus, after 2111 Twin Cams had been built, some 400 sets were still left.

There were also MGAs called de Luxe built in South Africa, which were not quite the same thing and are described in the chapter on production abroad.

Those chary of the Twin Cam's reputation for engine problems regard the de Luxe as the most desirable of all MGAs - the most powerful of the pushrod engines combined with disc brakes and the attractive Dunlop wheels.

Sports Car Graphic , which called it the Competition, remarked "While more expensive than the standard version, it provides a means by which the purchaser can end up with a competitive car at considerably less cost than that of converting the MGA and bringing it up to a similar state of preparation. For those interested in Class F production competition this latest offering from Abingdon is well worth the price of admission".

With a readily-available kit for the 1600 engine, the de Luxe could be tuned almost to Twin Cam performance levels. Alas, all too few have survived ...

New York distributor, J S Inskip, gave away this booklet, which runs over the next few pages, as a sales gimmick.

5

SUCCESS ABROAD

The North American story

When the MGA reached the United States for the first time, its price at the port of entry was $2195, an increase of $200 on the last TF (local taxes had to be added to this figure for the retail price). It undercut by some $300 the Triumph TR2 - advertised as giving "pulse-pounding championship speed" - and was considerably cheaper than the Jaguar XK140 roadster at $3469. The MGA coupé which eventually followed was $2750.

The roadster was an immediate success when it appeared first at the New York Motor Show in May 1956. John Thornley, who was there, reported 3600 orders and said the plant was being extended to cope with demand.

Nuffield executives had tested the feasibility of selling MGs in the States back in 1947 and had discovered a market for the TC, sending six over, which were quickly sold. But the fact that MGs sold in the United States in such huge numbers was due in part to the persistence of one man, Jocelyn Hambro, a scion of the City of London banking family.

Hambro had trained in New York with Brown Brothers, Harriman, the oldest private bank in the US, so was the obvious choice for the posting when his father, Olaf, decided to open up export markets there to attract dollars the war-torn British economy so badly needed. In 1947 the Government decreed that no less than 75 per cent of British auto production should be exported. This "export-or-die" policy persisted in the industry for years.

Jocelyn started selling British goods like pottery and motorcycles, but then, perhaps capitalising on GI experiences of the MG during and after the war, began to introduce the TC Midget to the towns and cities of Texas. The TC had been a big seller to American servicemen in Europe, particularly those in the army of occupation in Germany. The Woodhouse Garage in Cologne, run by Jack and Rita Woodhouse, was a major outlet for Nuffield's direct export service, and established the beachhead in North America which Hambro was to exploit.

As a publicity stunt, he gave the Mayor of Dallas a grand tour through the city in a TC; the mayor, unused to such rinky-dink machinery, remarked that the car was like a first cousin to a malted milk machine!

Jocelyn had disastrous results trying to sell kippers (they went off) and pots of honey (they exploded), but managed against all the odds to set up a distribution network for MGs. As the American car makers would originally not allow their distributors to handle imported cars, he chose people like the local jukebox salesman in Dallas and a Philco radio dealer in California.

By the time the MGA arrived, a total of 32,000 TC, TD and TF Midgets had been sold in North America over seven years and the original one-man-band had become the Hambro Automobile Corporation, of 27 West 57th Street, New York, which had a staff of 500 to handle, among other things, part of the BMC franchise. In a typical

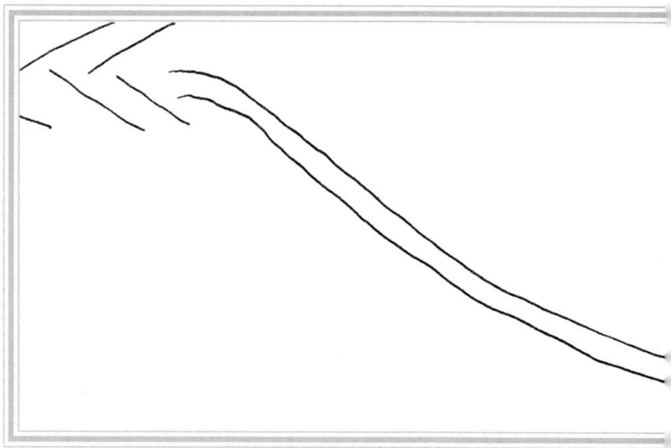

Part of a consignment of 25 white MGAs waiting to be loaded aboard the SS Emstein. They were destined for a Miami car rental firm.

financial year - to March 1959 - the company earned $53 million in car sales alone - much of this coming from 13,000 MGAs.

Hambro was essentially a financial and broking operation, arranging the dispatch of cars to distributors and remitting the proceeds to the United Kingdom. It was a complex set-up, to say the least. The Nuffield organisation required Abingdon to sell the car to Nuffield Exports, who sold it to Hambro's office in London. They moved the car to Chicago, where it was sold to a distributor, who sold it to a dealer who, in turn, sold it to a customer. Every hand the car passed though (except the customer's!) took a cut of the profit.

What is even more amazing is that Austin, another part of BMC, ran a parallel import operation of its own, so that Austin-Healey cars, eventually to be produced on neighbouring lines at Abingdon, were distributed quite separately at great cost, mirroring the highly-inefficient franchise operation

Windscreens and hoods were fitted at destination, so temporary covers were used during shipping.

Offloading in the United States in 1956. Some cars were crated by Nuffield exports before despatch.

in the UK. The situation was to improve in 1963 after BMC bought out Hambro and set up a joint company at 734 Grand Avenue, Ridgefield, New Jersey, handling cars, parts and service equipment for a network of 12 regional distributors and nearly 600 dealers.

By 1958, BMC was boasting it had sold more than 30,000 MGAs. "This model can claim to be one of the greatest, if not the greatest, single dollar earner produced in the United Kingdom. Its total earnings to date are in excess of 42 million dollars and are still rising fast". But this figure, of course, was turnover and not profit. Volatile exchange rates and intermittent deliveries at the time led Gus Ehrman, president of both Hambro and the New York distributor J S Inskip, to remark: "The only reliable thing about MG was that they were totally unreliable".

Inskip, the first and one of the largest of the distributors, took over the MG business from the multi-millionaire Collier brothers, Sam, Baron and Miles who had distributed a few MGs in the United States before and after the war, largely because of their own motor racing interests.

The Inskip headquarters at 304 East 64th Street also held the Rolls-Royce distributorship and it was from there that a network of dealers was set up in southern New England, New Jersey, Pennsylvania and Delaware which was to be so successful in selling the MGA.

Another of the major distributors was S H Arnolt in Chicago, widely known as "Wacky". From the heart of his empire at 415 East Erie Street, Wacky distributed to one dealer in Nebraska, two in Iowa, seven each in Indiana, Wisconsin and Michigan, 12 in Illinois and 13 in Ohio. Although he was an entrepreneur who also owned a Mexican coconut plantation and a tubular furniture firm, Arnolt was a sports car enthusiast. He became vice-president of the Italian coachbuilding company Bertone and put its bodies on 200 MG TD Midgets. He also raced Arnolt-Bristols at Le Mans and had a thriving business in Warsaw, Indiana, making accessories like heaters and mirrors for British sports cars, which their manufacturers had omitted to fit in the first place.

Other MGA distributors were Royston in Philadelphia; Ship and Shore, West Palm Beach, Florida; British Motor Car Distributors in Northern California and the Pacific North West (run by Kjell Kvale, who was later to play a part in the Jensen story) and Gough Industries in Los Angeles, who covered Southern California, Arizona, Utah and Nevada.

Gough introduced the new car to

One of the first BMC distributors in America was Waco Motors of Miami, here delivering four MGAs, a Morris Oxford and a Minor to Florida dealers.

its dealers and the press in the parking lot of the Santa Anita horse-racing track. Two small circuits were set up in the vast lot and more than a dozen MGAs were made available for time-trials.

Being an MG distributor was not unrewarding as their cut was 33.33 per cent of the retail price, compared with 20 per cent in the UK and far more than US manufacturers offered. This high mark-up was undoubtedly a result of Hambro's difficulties in setting up a distribution network from scratch. The handful of dealers the distributors initially serviced eventually became hundreds - including well-known and old-established names like Peter Satori in Pasadena, and Holiday Autos of Sherman Oaks, in California; Lawrence Falvery in Troy, Michigan and Ed Roth in Glassboro, New Jersey. By 1959, as the first MGA 1600 was being despatched, there were 600 franchised MG dealers in the US.

To put their importance into proportion, dealers sold around 19,000 MGs that year; of the other importers, Volkswagen sold 120,000 cars while the US domestic manufacturers mopped up 6.5 million sales.

While some dealers were punctilious in their service, others took on the MG franchise because their own domestic franchises were not successful - Studebaker and AMC dealers are often mentioned in this context - and they failed to give the spares and service required.

MG saw itself essentially as competing with other importers and not against Detroit's 'sporty' cars like the two-seater Ford Thunderbird from 1955, and the Chevrolet Corvette, which were not in the same league for responsiveness and handling. The Corvette was memorably described by style guru Stephen Bayley as having barge-like suspension which made it feel like a raft in a swell and "two-speed automatic transmission which responded to the driver's commands with the sensitivity of Hoover Dam sluices".

Though the North American market was vital to Abingdon, swallowing more than 80 per cent of MGA production, in US volume terms it was minuscule, and even that tiny market for imported sports cars was hotly contested. The first Triumph TR2 had appeared at the spring Motor Show in New York in 1954 and sold well, though not as successfully as the MGA.

In 1958 the Austin-Healey Sprite entered the market, priced at $1795 - well below the MGA at $2444 and the Austin-Healey 3000 which began at $3051 (without tax included). The intention was for the Sprite to complement the BMC range and not compete directly with the MGA. Research by Timothy Whisler, detailed in his book on British sports cars in America, *At the End of the Road,* shows that this strategy was mainly successful, since MGA sales were not diluted except in times of recession, when sporting drivers went for a cheaper option.

A more potent threat was the dart-shaped Sunbeam Alpine, launched in 1959 and aimed squarely at the North American market with prominent rear tailfins imitating Virgil Exner's 'for-

A 1600 at the New York Motor Show. Sydney Enever would have wanted to redesign the car on the right.

ward look' for the 1957 Chryslers. Compared with the ageing MGA (until then the best-selling imported sports car in America), and the Triumph, it was tighter, quieter and more comfortable. The top was easy to erect and it had wind-up windows. Ventilation was also better. One San Diego MGA owner wrote plaintively to *Autocar* in 1959: "The lack of cockpit ventilation, is in Californian weather, almost unforgiveable ... outside of that I love it madly!"

Doyen of auto testers, Tom McCahill of *Mechanix Illustrated* , reported on the Alpine: "At $2599 it's the best offering dollar-for-dollar since the

When the 1600 Mk II appeared in North America, much was made of the new recessed grille, set back, it was said, to avoid parking dents. This was not mentioned in European publicity.

Miss MG North America was voted in each year by Morris dealers and was, in 1961, "Canadian beauty red-haired Inge Britt". Her car not only had a make-up kit fitted in the door, but also a tasteful lace doily around the number plate ...

first Austin-Healey hit here in the early fifties. As a top rally car and possibly production class (SCCA) sports car it's going to take a lot to beat it".

That year of 1959 was the best for the MGA in North America; 19,591 models were sold. But the following year saw a steep decline to 13,084 and in 1961 the figure was down to 4210. In the MGA's final year of 1962, it was 2557 as the factory geared up to make its replacement, the MGB.

Americans were used to seeing changes in domestic car models every year; after six years, the love affair with the almost-unchanged MGA was beginning to wane, as it was for Triumph which fared even less well with unit sales plunging from 16,414 to 1285 by 1960, when the TR was being sold at cut prices.

When the MGA was first launched in the US *Road and Track* had asked: "How are they going to supply enough cars to meet the demand?" They probably did not in the early years, given the arcane and expensive system of distribution, but nevertheless the sale of more than 80,000 cars in such a tough market was a considerable achievement. It was helped greatly by the perception of the MGA as a 'lifestyle' object - an exciting car for the young; magazine adverts for products as diverse as shirts and life insurance featured glamorous young couples in MGAs.

This was in addition to Hambro's heavy advertising in motor publications. Some stressed the car's roadability, others capitalised on Sebring successes, but the lifestyle issue was always prominent.

Distributors like Gough Industries on the West Coast ran their own advertisements, emphasizing the fun to be had -

Heaven knows your new MG handles like a lamb. With nary a tire 'chirp' when you're teaching Detroit iron a little respect on a curve....we warrant there's nothing more fun-giving this side of Abingdon-on-Thames - for even thrice its modest tariff.

This was in November 1957 when roadster prices without tax and licence had gone up to $2445.

For North American owners used to Detroit iron, the fun was sometimes mixed with exasperation as they wrestled with the machinations of Joseph Lucas - often referred to as "the prince of darkness" - and his electrical equipment.

BMC had set up service schools in New York and San Francisco and even funded a mobile school to train mechanics in the strange ways of MGAs and other British cars. But lots of owners found themselves on their own when trouble struck.

Author P J O'Rourke, writing wryly in *Playboy* of his MGA experiences, claimed that they didn't understand oil at MG -

They forgot that oil is a liquid. And it dribbled out of all the holes with which the MG crankcase, transmission housing and differential were supplied for reasons of weight reduction. MG engineers even invented refillable shock absorbers so that oil could leak out of those too.

To an MGA owner these are mere foibles!

Canada

MGAs made for Canada were identical in specification to those intended for the United States, but took a different route to North America. They came by cargo ship from England down the St Lawrence Seaway into Lake Ontario, where they were unloaded at Hamilton. There was a small MG facility there, run from the main BMC base at Ottowa, where the cars were prepared for despatch to dealers, having windshields and lights fitted.

Those cars destined for western Canada were sent by train across to Winnipeg, from where the local distributor shipped them on to dealers on the west coast in British Columbia.

The main centres for MGAs were southern Ontario, the Maritime provinces and Vancouver. There were very few in central Canada where the weather is not kind to sports cars.

MG North American production figures do not differentiate between the US and Canada, but it is estimated that the total number sold in Canada was fewer than 1000.

Assembly abroad

Not all MGAs were assembled at Abingdon. Many were shipped out CKD (completely knocked-down) to be put

Pleasing the owner

Many MGA owners picked up their cars from the factory and none came as far as Harry Cape, an Australian sheep farmer from Merriwa, New South Wales, and an amateur race and rally driver. He ordered his Twin Cam car in May 1959 and then visited the factory to finalise the specification. What happened next was a demonstration of how far MG would go to please owners of its cars.

Harry had wanted a dark green car "to merge with the scenery so the cops won't see me", but all the ones on offer were red, white and the light colours preferred by North American customers. However, after personal representations to MG's boss, John Thornley, Harry's car was provided in the colour required, and also trimmed in non-standard dark green leather. Harry took delivery on July 6 1954. After a short continental tour, he took it back to Abingdon complaining that the engine was a bit rough. Thornley himself took it for a test drive and ordered the engine to be taken out and balanced.

When Harry arrived to pick up the car, Thornley asked him if there was anything else he would like done. Harry thought 2 inch carburettors might be better than $1^{3/4}$inch SUs, so out came the engine again for new porting of the manifolds. When the car finally left the development department on October 26 1959, the work had been completed; in addition there were five R5 racing tyres, an extra wheel and a 4.55 final drive ratio fitted. "It was done for the love of it" said Harry, quoted in *Wheel Spin*, the newsletter of the Victoria MG Car Club. "That's what the MG factory was like. It's a pity they're not still there, isn't it?"

The car is still in Australia and was last heard of in Tasmania.

Car identification plate from Australian-built 1600 Mk II. In the first line, 51 indicates made in Australia. Moss Green was not an MGA colour, but was listed in the BMC range, so this car would have been specially ordered.

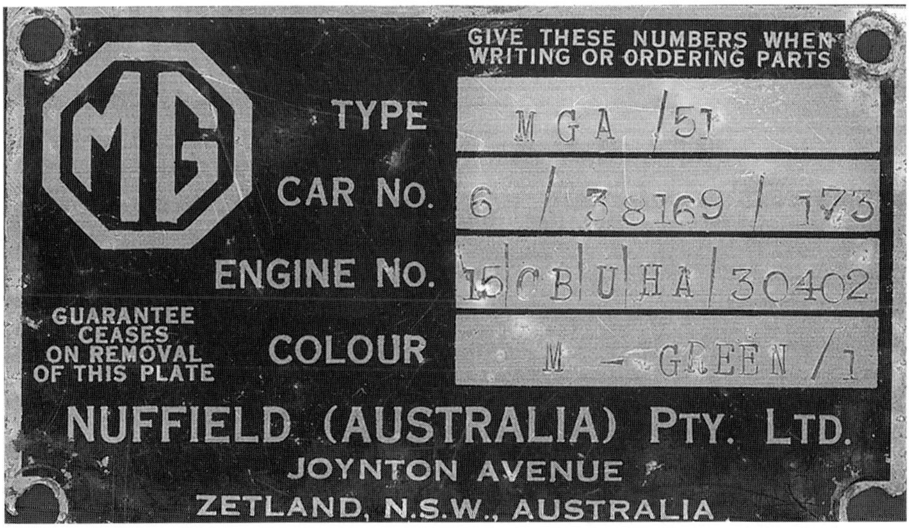

Hot work on the line at Enfield. All Australian-built cars had wire wheels.

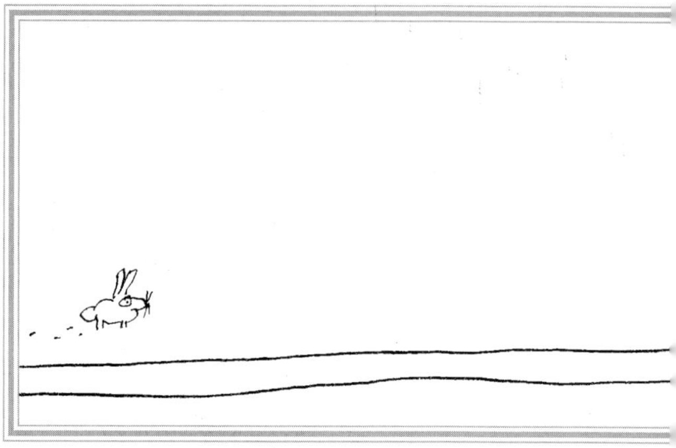

Extracts from the factory records show the wide variety of colours which were offered as standard. There were even options for door seal colours.

CARD_NO	ASSEMBLE	MODE	MODEL	DATE_OFF	DATE_O	CHASSIS_NO	BODY	ENGINE_NO	GEARB	BODY_COLOR	TRIM	DOOR_SE	TYRE_MAKE	PAINT_R	CARPE
55096	P.M.C	1957	SPORTS	6/23/58	6/24/58	43475	402	36375	1407	BARDIMAN GREY	CHERR	CHERRY	DUNLOP		BLACK
55823	P.M.C	1957	SPORTS	7/1/58	2/7/58	45646	428	33956	1474	RED	CHERR	CHERRY	GOODYEAR		BLACK
55849	P.M.C.	1957	SPORTS	7/11/58	7/17/58	46695	448	38025	1504	BARDIMAN GREY	RED	RED	DUNLOP		BLACK
55866	P.M.C.	1958	SPORTS	7/21/58	7/22/58	47137	472	33911	1499	BARDIMAN GREY	RED	RED	GOODYEAR		BLACK
63304	P.M.C.	1957	SPORTS	8/11/58	8/14/58	49107	510	37743	1554	BARDIMAN GREY	RED GR	RED	GOODYEAR		BLACK
63315	P.M.C.	1957	SPORTS	8/19/58	8/20/58	50285	518	42225	1576	BILL BLUE	TAN GR	TAN			BLACK
63376	P.M.C.	1958	SPORTS	10/9/58	10/13/58	51143	581	42211	1526	CARNATION RED	C.RED	C.RED	GOODYEAR		BLACK
63389	P.M.C.	1958	SPORTS	10/24/58	10/29/58	52310	591	93944	1630	BILL BLUE	TAN GR	TAN	GOODYEAR		BLACK
69534	P.M.C.	1959	SPORTS			53243	639	38032	1680	BARDIMAN GREY	C.RED	C.RED	GOODYEAR		BLACK
69585	P.M.C.	1959	SPORTS		11/11/11	54330	688	46167	1725	BARDIMAN GREY	C.RED	C.RED	DUNLOP		BLACK
76625	P.M.C.	1959	SPORTS		11/11/11	727	55107	46580	1773	BARDIMAN GREY	C.RED	C.RED	OLYMPIC		BLACK
76651	P.M.C.	1959	SPORTS			55974	754	46955	1794	BARDIMAN GREY	C.RED	C.RED	DUNLOP		BLACK
76671	P.M.C.	1959	SPORTS		4/29/59	7/56343/776	776	46968	1822	BARDIMAN GREY	C.RED	C.RED	DUNLOP		BLACK
76699	P.M.C.	1959	SPORTS			YGHN/2/8/56695	803	16GA/U/47293	1858	MAMBA BLACK	RED GR	RED	OLYMPIC		BLACK
81903	P.M.C.	1959	SPORTS			YGHN/2/8/56692	808	16GA/U/47260	1851	BILL BLUE	TAN GR	TAN	OLYMPIC		BLACK
81905	P.M.C.	1959	SPORTS			YGHN/2/8/56693	811	16GA/U/47291	1854	GUAN GREEN	TAN GR	TAN	OLYMPIC		BLACK
81907	P.M.C.	1959	SPORTS			YGHN/2/8/56697	812	16GA/U/47295	1856	GUAN GREEN	TAN GR	TAN	OLYMPIC		BLACK
81916	P.M.C.	1959	SPORTS		6/4/59	YGHN/2/8/56698	821	16GA/U/47296	1860	BARDIMAN GREY	RED GR	GREY	OLYMPIC		BLACK
81921	P.M.C.	1959	SPORTS			YGHN/2/8/56691	827	16GA/U/47259	1850	CARNATION RED	GREY	RED	GOODYEAR		BLACK
81920	P.M.C.	1959	SPORTS			YGHN/2/8/56694	823	16GA/U/47292	1855	BARDIMAN GREY	RED GR	RED	GOODYEAR		BLACK
76700	P.M.C.	1959	SPORTS		5/12/59	7/56696/806	806	47294	1857	MAMBA BLACK	RED GR	RED	GOODYEAR		BLACK
75367	BMC ENGLA	1959	COUPE			64748	0	4271		BLUE	GREY		DUNLOP		
81957	P.M.C.	1959	ROADSTER	7/14/59		65930	852	30437	1895	BARDIMAN GREY	RED	GREY	OLYMPIC		BLACK
81969	P.M.C.	1959	ROADSTER	7/27/59		66193	860	3658	1908	CARNATION RED	RED	RED	DUNLOP		BLACK
95102	P.M.C.	1959	ROADSTER	9/7/59	9/11/59	7/67070/895	895	6824	1937	BARDIMAN GREY	RED	GREY	GOODYEAR		BLACK
17770	UK-CBU	0	1600 COUPE		10/26/59	GHDL-73094	0	16A/U/4037		RED	BLACK	RED	DUNLOP		
95140	P.M.C.	1959	SPORTS	11/6/59	11/18/59	7/68907/508	508	16GA/U/139	5294	MAMBA BLACK	RED	GREY	OLYMPIC		BLACK
96177	P.M.C.	1960	SPORTS	12/1/59		YGHN/2/8/69673	536	16GA/U/2362	9924	BILL BLUE	TAN	GREY	DUNLOP		BLACK
95182	P.M.C.	1959	SPORTS	12/3/59		69663	541	2382	100	GUAN GREEN	TAN	GREY	OLYMPIC		BLACK
95183	P.M.C.	1959	SPORTS	12/3/59		YGHN/2/8/69678	542	16GA/U/2376	159	GUAN GREEN	TAN	GREY	OLYMPIC		BLACK
95181	P.M.C.	1959	SPORTS	12/3/59		YGHN/2/8/69679	539	16GA/U/2397	9629	GUAN GREEN	TAN	GREY	OLYMPIC		BLACK
95185	P.M.C.	1959	SPORTS	12/4/59		YGHN/2/8/69675	544	16GA/U/2433	9584	MAMBA BLACK	RED	GREY	OLYMPIC		BLACK
95195	P.M.C.	1959	SPORT/S	12/10/59		YGHN/2/8/69677	554	16GA/U/2428	2624	BARDIMAN GREY	RED	GREY	DUNLOP		BLACK
95192	P.M.C.	1959	SPORTS	12/9/59	12/14/59	YGHN/2/8/69676	551	16GA/U/944	9666	BARDIMAN GREY	RED	GREY	DUNLOP		BLACK
17779	UK-CBU-(BO	1959	1600		11/13/59	GHN/69761	0	16GA/U/1032		GREY	RED	RED			BLACK

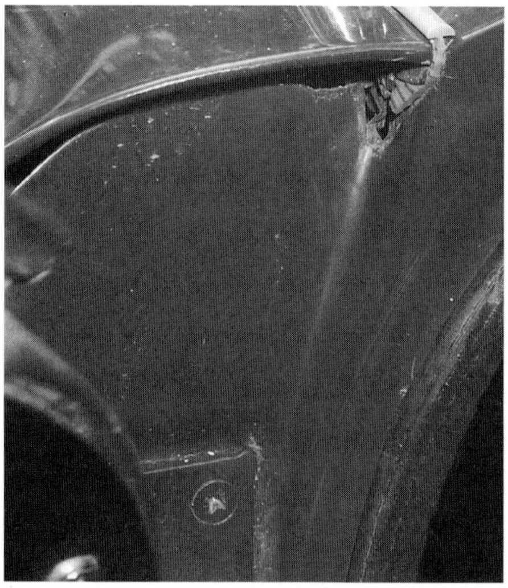

Sydney cars were finished differently at the top of the door jamb.

together abroad, in some cases because of import restrictions on fully-assembled cars. CKD in this context meant that the chassis, engine and all body parts arrived in boxes to be put together and painted in much the same way as they were at Abingdon. Exports began in early 1957.

Australia

Assembly of knocked-down MGAs in Australia began after the despatch from the UK in February 1957 of a batch of eight right-hand drive cars, chassis 25512 to 25519, which had actually been produced the year before.

MGAs were assembled by the Pressed Metal Corporation at their 20-acre plant at Enfield, near Sydney. It was a busy place; they were also putting together 32 other types of vehicle and the MGA was built in small numbers on the same track as 3-ton trucks and 1/2-ton vans.

In September 1957, BMC opened a huge engine and assembly plant at Victoria Park, Zetland, an old Sydney racecourse. It had a capacity of 1000 cars a week, producing the Wolseley 1500, the Austin Lancer and the Morris Major, with a high proportion of Australian-sourced components, including B-series engines. But the roadster continued to be produced at Enfield, using some engines and back axles from Zetland and others from the UK.

The Zetland works was subsequently demolished to make room for extensions to the runway at Sydney airport. There are apocryphal tales of Twin Cam engines being used for the landfill ...

Other local content in the roadster included: tyres and tubes (Dunlop, Goodyear, Olympic, Hardie) rear springs (marked "Made in Australia"), shock absorbers, floorboards, hoods, interior trim, coil and plugs. The cars all had vinyl rather than leather upholstery, black Westminster carpet, wire wheels, and a 4.3 rear axle ratio. Heaters were not fitted normally.

The MGAs began life in Australia at Sydney docks, where huge packing cases containing body parts, chassis and engines were unloaded and transferred to Enfield. The chassis went down the line upside-down, having steering and suspension parts attached and then turned over for the petrol pump and some of the wiring harness to be added.

Meanwhile, the unpainted body parts were welded together on a jig, the shell lead-loaded and painted, then carried on a slave chassis to the line to be dropped on to the chassis which, by now, had the engine in place.

Paintwork was in baked enamel, rather than cellulose, to cope with the Australian heat. The colours were different to those of the UK cars, and also offered something as standard that Abingdon did not - British Racing Green. Far more colours were offered than the standard British six or so, and they were exotic in name and tint:

1500
Alpine Green
Bardiman (Lark) Grey
Billeroy (Lightning) Blue
Carnation Red (engine red)
Connaught Green
Guan Green
Mamba Black
Nurburg White

1600
Araluen Blue
Hawthorn (British Racing Green)
Camden Cream
Cherry Red
Cheviot Fawn
Crocus Yellow
Cumulus Grey
Empire Green
English Grey
Gampian Grey
Inca Yellow
Kerrim (royal) Blue
Monza Red
Nurburg White

Two who did it the hard way ... David McMillan and Norman Tosh took their car to Australia when they emigrated there. Towing a trailer and having the spare wheel on the bonnet must have made the handling unusual to say the least.

The first Twin Cam coupé assembled in South Africa was delivered to racing driver Bob Olthoff. He modified it heavily and it was campaigned successfully in southern Africa before he brought it to England in 1960.

Orchid
Orient Red
Silverpine Green
Sky (Monaco) Blue
Spruce Green
Stradbroke Grey

After final inspection there was a four-mile road test, before delivery to dealers like Winterbottoms in Perth, Lanes Motors and Kellow Falkiners in Melbourne, P & R Williams in Sydney, Howards in Brisbane and All British Motors in Adelaide.

Only roadsters were made in Australia and they differed in several respects to the Abingdon-built cars and, in some ways, were better made. The outer sills, which were pop-riveted to the door pillars in Britain, were welded in Sydney; the spare wheel cover was attached to the back of the boot with nuts and bolts rather than rivets and there was much more lead-loading, particularly around the rear rail and the front door pillar, at the top of the front door jamb, where the UK-built cars had a cover plate. There were also some tiny detail differences, such as the routing of the rear tail light wire, and the coil of the oil pressure gauge line going under the bonnet rather than under the dash.

Engine blocks had the number stamped in rather than being on an aluminium plate; the chassis/engine number plate identified it as Australian-built and the chassis number itself was also stamped in the front off-side angle rail of the chassis rather than on the chassis crossmember.

Production was small, peaking at about 400 cars a year in 1957 and 1958. The total number of Australian-built cars was somewhere over 1000.

The factory record cards are held by the MG Centre of Sydney, 8, East Street, Granville, NSW, who can identify build dates and specifications.

South Africa

MGA parts were shipped to South Africa, minus components like tyres, windscreen glass and batteries. The first batch of eight, chassis numbers 24806 to 24813, was sent out to Durban in January 1957. The cars were painted green with black interiors.

Motor Assemblies Ltd, who put the kits together, had already successfully made 345 TD Midgets between 1950 and 1951. Work began in March 1957, and the production figures in the table below were supplied by Motor Assemblies in 1976 to the Cape Town centre of the MG Car Club.

There is one unusual fact here - the building, in December 1958, of what are described as four MGA sports de Luxe, followed by another 96 the following year. The UK de Luxe - a pushrod engine in a Twin Cam chassis - did not appear until June 1960 and used the 1600 engine. It was introduced the month after the Twin Cam engine had been abandoned to use up the remaining chassis. But the South African de Luxe was the standard pushrod car with the substitution of wire wheels and competition seats for disc wheels and normal seats.

Motor Assemblies were well used to turning out British sporting cars. At the same time that they were building the MGA they made 750 Magnettes between April 1957 and May 1959, and more than a thousand Triumph TR2 and TR3 models.

Europe

A handful of cars were assembled in Ireland by Booth Poole in Dublin and in the Netherlands by J J Molenaar at Amersfoot. The major locally-sourced components were usually tyres and batteries.

	1957	***1958***	***1959***	***1960***	***1961***	***1962***	***Total***
1500 sports	216	88					304
1500 sports de Luxe	4	96					100
1600 sports				76	56	12	144
1600 sports de Luxe			36	10	8		54
1600 coupé de Luxe			28	21	5		54
Twin Cam sports			52	4	12		68
Twin Cam coupé			16	4	8		28*
Total:							752

**British records put this figure at 24*

Russ Lawson's MGA in Sun City, South Africa.

Mexico City

Assembly by Automoviles Ingleses, beginning in February 1957 with a batch of five green cars, all with wire wheels and the 4.55:1 rear axle.

Cuba

One CKD Twin Cam was sent to Cuba in December 1958 for some enterprising owner to put together. It may be the car spotted in Havana in 1996 with a Lada engine ...

6

IN COMPETITION

Rally successes

In one of those sudden lurches of policy which characterise large organisations, BMC suddenly decided in 1954 that competition was a good thing after all. Abingdon, as explained in the chapter on Le Mans, became the official competition department headquarters.

But instead of concentrating efforts on the sporting cars which were likely to succeed, the corporation threw almost every existing model in the range on to the track and rally route. Peter Browning, who worked in the press office at Abingdon, puts it down to the need to appease all the conflicting interests in BMC at the time. In one season alone, in 1959, the factory entered in events the Austin A35, A40 and A105, the Healey 100/6 and Sprite, the Wolseley 1500 and 6/90, the Riley 1.5, the MGA Twin Cam, the Morris Minor 1000 and the Morris Mini-Minor.

John Sprinzel, who drove a Twin Cam in the Tulip Rally that year, remarked that there were so many models that no-one every really got to grips with any of them. He believed that had the factory adopted a 'one-model' policy for rallying and built up a team with expertise, the Twin Cam would have been an all-time outright winner. As it was, the MGA competed with some distinction in Europe in the early years of its production, though in later years, after the disappointments of the Twin Cam, the laurels tended to be taken by Austin Healey. As Peter Browning says: "In a rally there's no substitute for sheer power", something the Healey had in abundance.

So, in Europe, the MGA was targeted at team prizes, class wins and ladies' cups. In the United States, however, it continued to do well in the Sebring endurance race.

The MGA has a long history of competition over the last few decades, and there are several excellent books which cover it, so what follows is a summary of its major successes while the car was still in production.

The first major European success after Le Mans was the tough Mille Miglia in 1956. It was a race which MG had not contested since 1933, when the Magnette K3 took the 1100cc class prize. The newly-formed competitions department, run by Marcus Chambers, was persuaded to enter two cars. Officially, after the tragedies in 1955 at Le Mans and Dundrod, BMC did not take part in racing, but this was a road rather than a track event. In April 1956, for the 23rd Mille Miglia - not quite a thousand, but 991 miles - there were 370 entries, including seven MGs, two of them factory-sponsored MGAs. None of these was going to win outright against the Ferraris on their home ground - in fact, the Modena cars did the course in some 11 hours, taking the first five places.

From the start at Brescia to its finish there, the event was dogged by incessant rain and bitter cold. The works MGs were roadsters with aero screens - the worst possible combination for the weather - and the water poured into them. They were driven by Nancy Mitchell with Pat Faichney as navigator and Peter Scott-Russell with Abingdon's test driver Tom Haig. The

The team for the 1956 Mille Miglia - Peter Scott-Russell and Tom Haig with Pat Faichney and Nancy Mitchell in 'Mabel'.

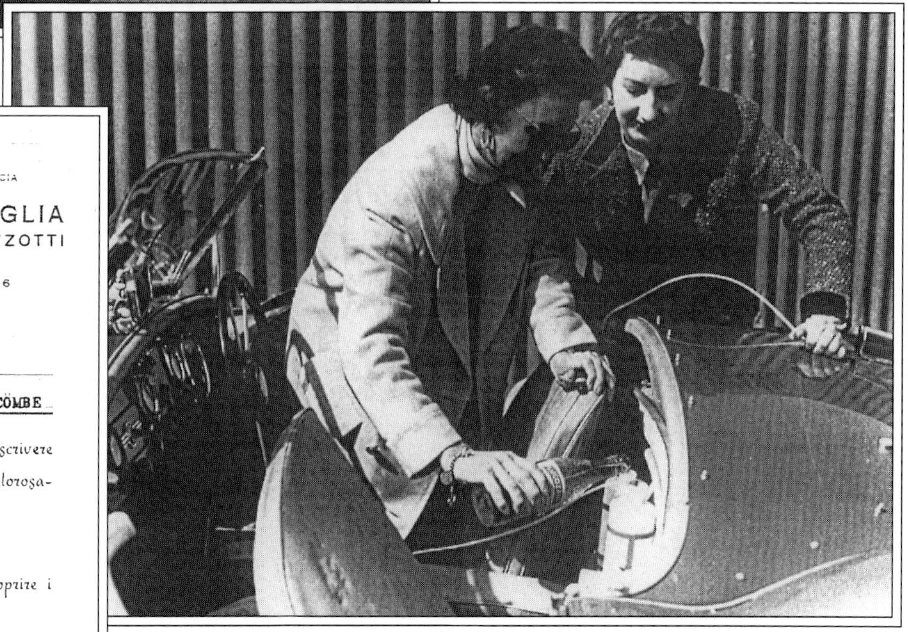

Windscreen washer bottles filled with Lucozade kept them going.

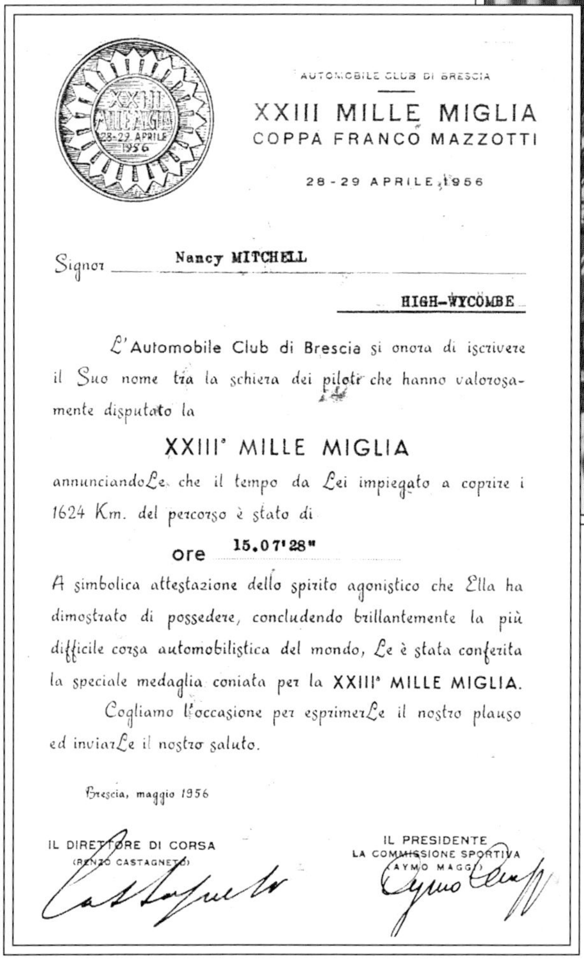

AUTOMOBILE CLUB DI BRESCIA

XXIII MILLE MIGLIA
COPPA FRANCO MAZZOTTI

28 - 29 APRILE 1956

Signor Nancy MITCHELL

HIGH-WYCOMBE

L'Automobile Club di Brescia si onora di iscrivere il Suo nome tra la schiera dei piloti che hanno valorosamente disputato la

XXIII° MILLE MIGLIA

annunciandoLe che il tempo da Lei impiegato a coprire i 1624 Km. del percorso è stato di

ore 15.07'28"

A simbolica attestazione dello spirito agonistico che Ella ha dimostrato di possedere, concludendo brillantemente la più difficile corsa automobilistica del mondo, Le è stata conferita la speciale medaglia coniata per la XXIII° MILLE MIGLIA.

Cogliamo l'occasione per esprimerLe il nostro plauso ed inviarLe il nostro saluto.

Brescia, maggio 1956

IL DIRETTORE DI CORSA (RENZO CASTAGNETO)

IL PRESIDENTE LA COMMISSIONE SPORTIVA (AYMO MAGGI)

A thousand miles in 15 hours - "finishing brilliantly the most difficult motor race in the world" says the citation.

Mabel and Nancy, with a Miss Reece, took another Ladies' Cup in the 1957 Lyons-Charbonniéres rally.

cars were tuned to give about 10bhp more than the standard road car; they had Le Mans-style petrol filler caps, 20-gallon long-range fuel tanks, high ratio rear axles, Andrex friction dampers and oil coolers. They were painted Alfa Red, which came in useful as Scott-Russell noticed that Italian crossing keepers would lift the barriers for cars in their national racing colour, but leave them down for others. Chambers, following in an Austin Westminster, took the decision then that the works team would always have red cars with white tops.

After an horrendous journey through storm and flood down the Adriatic coast to Pescara, across to Rome then up through Florence and Bologna, Scott-Russell, in MJB 167, arrived back in Brescia 15 hours, 2 minutes and 15 seconds after setting out. Nancy Mitchell, in her famous 'Mabel' MBL 867, was some five minutes behind him and was so exhausted that she had to be lifted out of the car at the finish, having driven the entire route at an average of more than 65mph to win the ladies' prize.

The two teams were second and third in the up to 2-litre limited price class and in the up to 1500cc class, MGAs came fourth, fifth and sixth. It was an event again marred by tragedy; five people were killed and 20 injured as the cars skidded off greasy roads into the crowds.

The two cars were driven back to England by Scott-Russell and Haig who, when they got to London, telephoned John Thornley and suggested that the muddy, battered cars be displayed in the BMC showroom in the West End. Thornley vetoed the idea immediately: "Len Lord doesn't even know we had two MGAs in the Mille Miglia ... bring them back to Abingdon and we'll put a cutting torch through them!" Luckily this did not happen, and Mabel has been restored by MG expert, Bob West.

Nancy Mitchell, described in the newspapers as 'housewife and mother of three', was a perfectionist who, to avoid spending extra vital seconds at

Where did she keep them all? Some of Nancy Mitchell's trophies.

Nancy and Mabel, reunited at an MG Car Club event at Abingdon in 1992, with Bob West who restored and now owns the car. Nancy Mitchell died in 1996 at the age of 81.

Below: MGAs had some success in the tough Rome-Liégé-Rome rally, but Belgians Harris and Hacquin, seen here at Merano in 1957, were knocked out by suspension problems.

necessary roadside stops, even wore special loose cami-knickers. She was to do even better in the next event, the Alpine Rally, partnered again by 'hotelier's daughter', Pat Faichney. They took Abingdon's first Coupe des Alpes (awarded for completing the course without penalties) and the Coupe des Dames.

Five MGAs started from Marseilles. This time they were hardtops, to get around the French ban on open cars in rallies. They had wire wheels, maximum rpm tell-tales on the rev counters and large fuel tanks. They were originally fitted with oil coolers but these were removed because of faulty unions and they ran without trouble on Castrol R.

Of the 79 cars from seven nations which started, 34 finished, 17 of them qualifying for Coupes des Alpes.

All five MGAs, well supported by the works mechanics, ran strongly until the final stage. When one of the cars broke a rear spring, the mechanics changed it for a new one in 26.5 minutes. The mechanics were unable to rescue John Gott's car, which went out with a broken half-shaft, and the car driven by Jack Sears took off and overturned, smashing the hardtop but not injuring the crew.

Final MGA placings, other than the cups were third, fourth and fifth in the GT category.

The next major international event was the hardest rally in Europe, the Rome-Liége-Rome, a 90-hour marathon with no rest stops and a set average speed of 60kph to be maintained. Its title was a misnomer, since it started and finished at Spa in Belgium and went to Zagreb instead of Rome. Of the 86 starters, only 35 finished, three being works MGAs.

The BMC team, using the MGAs from the Alpine, was: John Gott and Chris Tooley who came 13th; Nancy Mitchell and Anne Hall (26th); Dick Benstead-Smith and John Milne (14th and best newcomers) and Gerry Burgess and Sam Croft-Pearson (out with a damaged sump). The women were runners-up for the ladies' prize, beaten by a French team in a Peugeot, and the team prize was snatched by Volvo, because two of the MGA crews spent too long in a support station, were caught by fog and were unable to make

The Morley twins, Donald and Erle, took their 1600 de Luxe to the team prize in the 1962 Monte Carlo Rally. Here, one of the twins navigates the hairpin on the way down to the seafront at Monaco.

The car, which has now been restored, also won the Tulip Rally.

up the lost time.

The following year - 1957 - was a bad one for rallying. A combination of the Suez crisis and previous year's tragedy at the Mille Miglia led to the cancellation of the Monte Carlo, Alpine and RAC rallies. But the 3000 mile Rome-Liége-Rome was run with MGAs and much the same team as the previous year. Nancy Mitchell took the ladies' prize, as she did in the Lyons-

Charbonniéres rally which, with her other rally successes in other BMC cars, gave her the Ladies' European Touring Championship two years running.

Chris Tooley, who drove the Rome-Liége-Rome with John Gott in MJB 167 and finished ninth overall, recalled that on the long mountain stages the car was hardly ever out of first or second gear and revs rarely dropped below 5000. But the car ran faultlessly from start to finish.

The big Healeys were now favoured by BMC for continental rallies. In the 1958 Liége there were four 100/6s and only one MGA; John Gott and Ray Brookes in a Twin Cam. Because of factory delays this MGA was almost straight off the production line and was the first Twin Cam to compete in an international rally. The instruction from team manager Marcus Chambers was to finish at all costs. Despite awful warnings about taking an untried car on such a gruelling event, the pair finished ninth and would have come higher but for a distributor fault on two of the main stages. Gott remarked afterwards that it was a real tonic to drive a 1600cc car which could keep up with the quickest Porsches.

The last major international successes while the MGA was still in production were in the Monte Carlo and Tulip rallies in 1962, when they finished first in class. In the Monte, the Morley twin's success was in a 1600 Mark II de Luxe coupé, which was also driven by Rauno Aaltonen and Gunnar Palm in the Tulip. The car was pressed into service yet again by John Gott for the Liége-Sofia-Liége, but without success.

Another restored works competition car is the one driven by John Gott in 1959 Monte. This rear view shows the large fuel tank.

By now Austin-Healeys and Minis were scooping most of the awards for BMC internationally. But there were, of course, many sporting successes at home over the years; in the RAC Rally, at Silverstone, at Goodwood. Private entrants like the Fitzwilliam racing team, which had considerable factory help, even to the extent of 'borrowing' factory cars, competed strongly at home and abroad in events like the Mille Miglia and the Nürburgring 1000 kilometre.

Privateers were, to say the very least, enterprising. When Fitzwilliam crashed car LJB 205 into a Scotch whisky advertisment in the 1957 Mille Miglia, he nursed the battered remains back to Brescia, abandoned the car and went on to the Nürburging to prepare for the following week's race. His wife, who had never before driven on the continent and had broken her back in a car accident three years previously, then drove the MGA back to Abingdon, where it was fitted with a new chassis and engine. On its first test run on the Oxford bypass, it crashed again and another race car had to be prepared for Fitwilliam and taken to Germany in time for him to compete at the Nürburgring

The Fitzwilliam Mille Miglia team cars in the garage at Brescia. Note the competition windsreens and racing-style filler caps. LBL 301 was a Le Mans car and LJB 205 the first production car, though registrations were often swapped around.

The Dick Jacobs Twin Cams in action at Silverstone during the BRDC International Trophy in 1959. Roy Bloxham, who won, leads Alan Foster. (Courtesy Castrol Ltd)

Below: Dicing in the wet at Brands. Removing radiator slats was a popular way of avoiding overheating.

Dick Jacobs, who had competed at Le Mans, ran the famous Mill Lane MG dealership in South Woodford, Essex until the M11 motorway was driven through it. Though he himself did not race again after his accident at Le Mans, he prepared MGAs and Twin Cams for others to compete in successfully. The Jacobs Twin Cams 1 MTW and 2 MTW which were such a sight at club meetings are still racing. Private owners are still entering the Monte and other major rallies

MGAs have never stopped competing and never will.

Competing in North America

Sebring is a track which will always be associated with the MGA, for it was here, within weeks of its launch in

Chris Tooley and Mike Reid in 1500s at an MG Car Club meeting at Silverstone in May 1959.

North America, that the car enjoyed its first major competition success.

The Automobile Racing Club of Florida, which had set out the 5.2-mile track on Hendryk's airfield, a wartime B17 training base abandoned by the US Air Force in 1946, advertised it as America's greatest endurance race for sports cars - a marathon in March of every year running from 10 in the morning until 10 at night which attracted all the big names in motor sport at the time - Moss, Fangio, Behra, Collins and others. From small beginnings in 1950, it had become the home of America's premier sports car event, where success also contributed towards the world sports car manufacturers' championship. The perfectly flat track, some of it on the old concrete runways, cut through the sandy brush of Highlands County in central Florida.

The American Automobile Association had withdrawn its support for the event in 1956 because of the fatalities at Le Mans a few months previously, but the FIA did not demur and it went ahead. Three MGAs were prepared by East Coast MGA agents and entered in the 1100-1500cc class. There was no official factory participation, except help with spares and assistance from the US importer Hambro. Hambro usually chose the Sebring drivers, all of whom, for this first event, had extensive experience in racing G T-series cars.

The New York distributor, Inskip, prepared car 49 for Gus Ehrman and Dave Ash (Ash went on to tour the East Coast signing up MG dealers). Manhattan Auto of Fairfax, near Washington DC, had Bill Kinchloe and Steve Spitler as drivers in car 50 and the third MGA, car 51 from Warren, Pennsylvania, was driven by Fred Allen and Johnny Van Driel.

The cars were standard MGAs, the engines of which had been stripped and rebuilt. Since each had been prepared separately there were slight differences, but all had extensive safety equipment and straight-through exhausts.

Opera star James Melton flagged the 60 drivers away to a Le-Mans style running start at 10am. The MG strategy was to lap at around 4mins 30secs, which the cars achieved fairly consistently; mechanical problems were few - a loose throttle linkage and heavy brake wear. The only mishap was when car 51 spun into a sandbank and lost a lot of time being dug out. The other two cars averaged 65.43mph and a well-driven race ended with the three MGAs crossing the line together in fourth,

Sebring in the fifties still had the atmosphere of a country track. Here, in 1957, the Fred Allen/Johnny Van Driel car dices with a Lotus 11.

The Sebring cars were prepared in the same way in both 1956 and 1957, though they were, in fact, different cars.

fifth and sixth places in class behind the Porsches. It was, in fact, a leisurely finish, since the MGAs were assured of the team prize before the race ended. In overall terms, car 50 was 19th, 49 was 20th and the sandbank victim, which completed only 139 laps, came in 22nd.

It was a notable success for the new car, much-trumpeted by BMC even though the Porsche Spyders could not be beaten, and the teams returned the following year with new cars and enjoyed even greater success. This time, they took not only the team prize but the Spitler car was first in class followed by Ehrman and Ash, with two Alfas close behind. The MGAs did far better than the big Healeys, which appeared at Sebring that year for the first time, but all were dogged by cracked conrods: only one finished - in 26th place.

The following year, 1958, no MGs were entered, but Healey saved its reputation by winning the team prize.

BMC returned officially to racing at Sebring in 1959, fielding three Twin Cams and a practice car, all prepared at Abingdon. The four, whose chassis numbers were 931 to 934, were originally left-hand drive roadsters when they were taken off the production line in February 1959. They were converted to right-hand drive coupés in the competitions department. One of the reasons for running them there was to prove the reliability of the Twin Cam to sceptical American buyers - bad reputation had spread ...

The rest of the BMC team comprised three Austin-Healey Sprites, which were prepared at the Healey headquarters at Warwick. To maintain the six race cars at Sebring there were 14 Canadian mechanics. The drivers also numbered 14: two for each car with two spare. The MGA drivers were: 28 Gus Ehrman/Ray Saidel; 29 Jim Parkinson/John Dalton; 30 Ray Pickering/Jack Flaherty.

The spare MGA driver was Sherm Decker of Oneonta, New York, well-known in competition circles and a noted MGA winner at Watkins Glen and Lime Rock. He was in charge of the workshop in the Gordie Morris dealership in Maryland, New York and an expert on Twin Cams. He maintained that the way to keep them sweet was to keep to the factory specifications, but increase the piston clearance. He was relegated to spare driver because of the Hambro method of apportioning drivers among big regional dealerships, but was to play a heroic role in getting number 30 to the finish.

Practice runs revealed - though no-one should have been surprised with temperatures regularly in the 80s - that heat build-up in the coupés was substantial and a menace to the drivers over a 12-hour period. The work-around was to drill holes near the pedals to admit fresh air, but when the race started it rained heavily, letting large amounts of water in from the saturated track and drenching the drivers' trousers.

Car 28 ran well apart from hitting a bale of straw and losing time and car 29 suffered a sticking starter motor which took time to repair.

As the race wore on and the Porsches gained ground, Decker was sent out in car 30 as he had proved very fast in practice. But 10 hours in, the Twin Cam engine blew. Decker pushed the car for a mile and a half in the dark to the finish line, with the other race cars narrowly avoiding him, and qualified as a finisher at 45th overall.

The two other MGs finished second and third in class after a Porsche. The Sprites did even better, finishing in the first three places in the 750-1000cc class and winning the manufacturer's team prize.

MG's boss, John Thornley, had watched the race and its preparations and determined that the team would

Canadians Ed Leavens and Fred Hayes in the pits at Sebring in 1960. The Twin Cam roadsters had larger wheelarches than standard cars.

Below: De Luxe coupés that ran at Sebring in 1961 and 1962 were tested at Silverstone before being despatched to Florida.

do even better the following year, 1960, when, again, three Twin Cams were entered. Healey increased its presence with three 3000s and two Sprites.

Records show that two of the MGAs, left-hand drive chassis 2571 and 2575, were converted to right-hand drive, finished in British Racing Green and completed in February 1960. This time they were fitted with aluminium hardtops to save weight and had better seats and instruments: a six-inch chronometric rev counter, separate switches for the twin fog lamps, Austin-Healey push-button starter and two-speed wipers to cope with rain.

The standard Twin Cam body was altered in several places. The boot floor was cut away to accommodate a 20-gallon fuel tank, and the rear of the cockpit was cut away and a raised glass-fibre panel fitted to allow for a peculiar Sebring requirement - an overnight suitcase. The wheelarches were enlarged and, learning from the previous year, a cool air vent in the bonnet directed air into the cockpit.

The engine was standard, blueprinted and ran with a 9.9 to 1 compression ratio. There was a close-ratio gearbox and 4.55:1 rear axle.

The MGA line-up was: 38 Ted Lund/Colin Escott (UK); 39 Ed Leavens/Fred Hayes (Canada); 40 Jim Parkinson/Jack Flaherty (US).

Alas, 38 was out within three laps with a broken valve spring and the two other cars suffered from fractured brake lines, finishing third and fourth in class at 69.1mph and 64.08mph respectively. The big Healeys had gearbox problems, though the Sprites again did well.

The MGAs had done well, but not well enough. However, the best was yet to come.

By the time Sebring came around the following year Twin Cam production had ceased, so the MG team fielded two 1600 de Luxe coupés - the Twin Cam chassis with a pushrod engine.

In January 1961, as the race plans were being laid at Abingdon, BMC's new 16GC engine of 1622cc was already available as it was being produced to go in the MGA 1600 Mark II in a few weeks' time. But to keep the cars in the under-1600cc class at Sebring, where they had the best chance of success, a special 1588cc block without water holes was cast to take the cylinder head, which had larger inlet

Under the famous MG bridge at Sebring goes the Parkinson/Flaherty car, which finished fifth in class.

In the 1961 contest, in which Porsche did not take part, the MGAs had a new rival, the Sunbeam Alpine, introduced in America in 1959 and successful in SCCA class G races from 1960. It had also given the MGA a run for its money at Silverstone.

The MGA team was: 43 Peter Riley/John Whitmore (UK); 44 Jim Parkinson/Jack Flaherty (US).

The spare driver was South African Bob Olthoff, who had arrived in England with MGA Twin Cam YRX 310 in 1960 and achieved considerable success in club meetings.

There were three Alpines, which were allowed to take an early lead as

and exhaust valves, from the higher-capacity engine. The crankshaft and main bearings were also from the larger engine. Syd Enever's instructions to the development department, dated January 9 1961, list some 50 changes to standard engine practice and instructions to build three engines.

From the Twin Cam came con rods, big-end bearings, modified SU $1^{3/4}$in carburettors and a close-ratio gearbox. Competition clutch and pistons were specified and two different types of camshaft, with different profiles at which valve crash occurred at around 6700rpm, were tried. Even the oil was specified - Castrol XL 30 - and modifications made so that oil pressure at 5500rpm was 70psi.

The cars were shipped to New York on the Queen Elizabeth, then transported by road to Florida.

Club racing with MGAs flourished in North America. Here, Dan Rubin's white roadster cuts up another MGA at Cumberland.

MG assessed the race. But then the MGAs began to gain the advantage because of their larger wheels (which meant tyres had to be changed only once) and much faster pit stops than the Sunbeam team could manage. At one stage towards the end of the race, the MGs were a lap ahead. At the end of the race, the MGAs were first and second in class. Car 44 finished 14th overall with 175 laps and car 43 was 16th with 173. Three Alpines occupied the next places. The lead Alpine finished 17th overall, behind the MGAs and a Sprite.

The Alpine team had done quite creditably, but Sunbeam's cheeky advertising made it into a famous victory. "... it proved its stamina and reliability with an outstanding performance ... that makes the budget-priced Alpine the equivalent of a middleweight champion. It beat every car in its price class." The Sunbeam sold at $2595 and the MGA at $2440. The same price class - or not?

The next year - and the last at Sebring for the MGA - there were four Alpines, four Porsches and three TVRs as competition. The MG team, in three 1600 de Luxe coupés, competed as Ecurie Safety Fast, capitalising on the motto on the MG bridge which crossed the middle of the circuit: 51 Jim Parkinson/Jack Flaherty (US); 52 Andrew Hedges/Jack Sears (UK); 53 John Whitmore/Bob Olthoff (UK/SA); Spare F Morrell.

Again, the obsolete 1588cc engine was used in order to be able to compete in the 1600GT class. The only distinguishing difference to the previous year's cars was in the front grille, where alternate slats were removed to give better airflow. The grille itself was the old type, not the new recessed variety.

Porsche returned to Sebring in triumph, taking the first two places: a Sunbeam Alpine was third. Of the MGAs, the Hedges/Sears car was fourth, Parkinson and Flaherty fifth and the other car sixth.

By the time Sebring came around the following year, 1962, MGA production was tailing off as the factory tooled up for the MGB.

A well-prepared race car; note the rear headrest and special fairing.

BMC's briefing to MGA dealers boasted of "consistent successes at Sebring", and while the wins were not achieved every year in the six years of competition, the return to endurance racing meant valuable lessons for MG and a high publicity profile.

It was company policy to allow the Sebring cars to remain in North America rather than return them to the UK, so several are still in private ownership there.

Record breakers

MG had been in the business of record-breaking since 1930, shortly after the first car was made. The company was constantly in the public eye before the second war, adding lustre to its sporting credentials by shattering records in the International Class H for cars of less than 750cc and building the first car of this size to reach 100mph. Names like Goldie Gardner and George Eyston became forever associated with the MG speed legend. The cars were given EX (experiment) numbers.

Syd Enever, the guiding light behind the MGA, had been involved with record-breaking since before Gardner won the Seagrave Trophy on the Frankfurt *autobahn* in 1938 in EX 135, with a one-way run of 196mph in a Class G car (up to 1100cc). The following year at Dessau in Germany, with a slightly larger engine taking him into Class F (up to 1500cc), Gardner reached 203mph. By 1950, after a break during the war, EX 135 held six international class records with engines ranging from 350 to 2000cc.

The significance to the MGA story

Ross St Croix, in a Twin Cam, finishing third at Mosport Park, Canada in the 200 production sportscar event. (Courtesy Castrol Ltd)

of these record attempts is that when the unsuccessful prototype MGA roadster HMO 6 was built in 1952, Enever had ordered a spare chassis. This foresight was rewarded two years later when the BMC board, at the prompting of George Eyston, decided to try to boost flagging T-series sales in the United States with a record attempt on the Bonneville salt flats in Utah.

HMO 6 was dragged out and fitted with an undershield, wheel spats, a cockpit cover and Perspex bubble, and put through wind tunnel trials by Armstrong-Siddeley's helpful aerodynamics expert, who happened to work at Abingdon with Syd Enever during the war. The results were not good enough; a new body was needed, which Enever was quick to design.

The spare chassis, which had been lying around in the experimental department, was perforated for lightness, probably by Henry Stone, described by chassis designer Terry Mitchell as "the master of hole-drilling" - it was said that if you left your spanners lying around in the workshop Henry would drill a neat set of holes in them for you! It was fitted with a tubular framework to which 18-gauge aluminium body panels were attached.

The engine was basically the 1466cc XPEG used in the TF Midget. Two versions were made; one for long-distance running and the other for the sprint. Both were heavily-tuned.

The car, now named EX 179, was shipped across the Atlantic and taken by rail to Utah, where it was to be driven by American Ken Miles and George Eyston, who was now 57 and had a 23-year record-breaking association with MG.

On August 17 1954, the record attempt began at dawn on the Flats, 120 miles west of Salt Lake City and just off Highway 40. There were two courses on the dried-up lake bed: a straight of more than 20 miles long with a measured mile and kilometre in the middle, and a circular track, 10 miles in diameter. The salt was smoothed by a truck dragging a heavy piece of iron behind it and the driving line marked with a six-inch wide band of crude oil.

The aim was to maintain an average of 120mph for 12 hours, stopping only for driver changes every three hours. The car ran flawlessly, lapping at around 125mph, taking 7 international and 28 American Class F records. Three days later, with the sprint engine and a higher axle ratio, Miles demolished the flying 10-mile records, hitting more than 153mph. For an unblown 1500cc pushrod engine these were astounding figures, confirming MG's reputation for speed and safety.

Much of the point of the exercise had been to promote the new 1500 engine and the fact that it was now fitted in the TF. But because of an over-supply of TF cars with the previous engine, it largely failed as a marketing initiative. Nevertheless, the achievement was used in the first brochure for the MGA, which pointed out that years of record-breaking experience had taught MG engineers how to design a

The EX 179 record-breaker, which reached 170mph, with Henry Stone, H Riddell and Cliff Bray who built it in the experimental department at Abingdon.

MGA on the Salt

How fast would an ordinary MGA go on the salt at Bonneville? Jim Frostrom, a judge in the 1959 National Speed Trials, took his 1500 roadster up to the maximum to find out.

The car was standard, other than that a straight-through exhaust was used and the windscreen was removed. With tyres at 44psi, well above the recommended maximum of 27 front and 31 rear, he reported in *Road & Track* a fastest run of 101.1mph. This does not seem particularly fast, but it was done at an altitude of more than 4000 feet.

production car of small frontal area and low overall drag.

In 1956, the flats were reserved between August 10th and 24th for another attempt by EX 179, this time with an early version of the twin cam engine which had been seen at the TT at Dundrod in one of the Le Mans cars.

The MGA chassis and B-series block were mated for the first time in a record attempt. The engine was fitted with two twin choke carburettors and deviated little in terms of power from the production version which was to follow, developing some 100bhp. With a special high-ratio rear axle, it gave 143mph at 5000rpm. Brakes, wheels, suspension and steering were standard. One of the few changes made, other than the engine, was conversion to right-hand drive.

In the hands of Ken Miles and Johnny Lockett, EX 179 took 16 international Class F and 46 American records, including the 12-hour record, circling the 10-mile track on the salt at 141.71mph. With the sprint version of the engine, the car achieved a new record of 170.15mph. This was impressive, but not enough for MG's management who, by now, had its hopes set firmly on 240mph - four miles a minute - which would also break Gardner's pre-war record of 203.9mph. For this, a completely new shape and configuration was required, with the driver reclining in the nose of the car and the engine between him and the rear axle.

EX 181, "the roaring raindrop" as it was known because of its beautiful aluminium shape, became the world's fastest 1500cc car, propelled by a supercharged twin cam engine. Terry Mitchell, who was intimately involved in the chassis and detail design, recalled how it began: "Syd said to me one day 'we've got to do another record car - you'll do it won't you?' I knew he'd had some quarter-scale and eighth-scale models made of various shapes and I got some aerofoil shapes out and said to Syd 'we've got to decide what shape we want'. He'd say 'yes' then buzz off!

"It got to the stage where they wanted the car within the year, so Syd came in one Saturday morning and decided on the shape. I drew it quarter-scale then had to scale it up full-size on these six-by-three plywood sheets for the bucks. Harry Herring (MG's model maker) did the wooden mockup and the body was built by Midland Sheet Metal at Nuneaton.

"As for the chassis, it was tubular and completely new. I got inspired. I had this bee in my bonnet about the chassis being like a tree, with branches growing out and tapering - trees stand up in the wind so the chassis would be all right".

Extensive wind-tunnel tests of at least six shapes at plane-makers Armstrong-Whitworth at Coventry resulted in almost the perfect aerofoil, with a drag factor of 0.000292, about a tenth of that of a normal sports car.

The twin cam power unit, made by Eddie Maher's team at Morris Engines, was again based directly on the B-series in the MGA and similar to the unit which was to power the Twin Cam itself, but was blown with a huge Shorrock supercharger, originally designed for a 10-litre diesel truck. This tripled engine power to some 300bhp and the block was stiffened to cope with the stresses.

A Riley RMA gearbox transmitted the power to the de Dion rear axle. Front suspension was from the MGA. The car was not designed to be braked quickly from high speed and had only one Girling disc brake at the rear to bring it to a standstill.

The shell and mechanicals were put together at Abingdon by Alec Hounslow.

Before it was shipped to Utah, MG test driver Tom Haig took the car through its paces on February 5 1957 at a US airforce base in England. Terry Mitchell takes up the story:

"One Saturday John Thornley had arranged for Tom, Syd and myself to go to Brize Norton, where the runway was three miles long. Tom was running the car up and down and the runway was so wide he could turn at 60mph and so clean that you could eat your food off any of the aircraft pans.

"A big American bomber came over and started circling. Syd asked the base commander who was with us, if we should get the car off to let him land. He replied 'Oh, he can circle up there, he's only come from the Azores'. He was more interested in our car than his plane".

The car was shipped to Utah and on August 20 1957 Stirling Moss arrived at Bonneville, having won the Grand Prix at Pescara two days before. The fastest he had driven before was at about 200mph in a BRM at Monza.

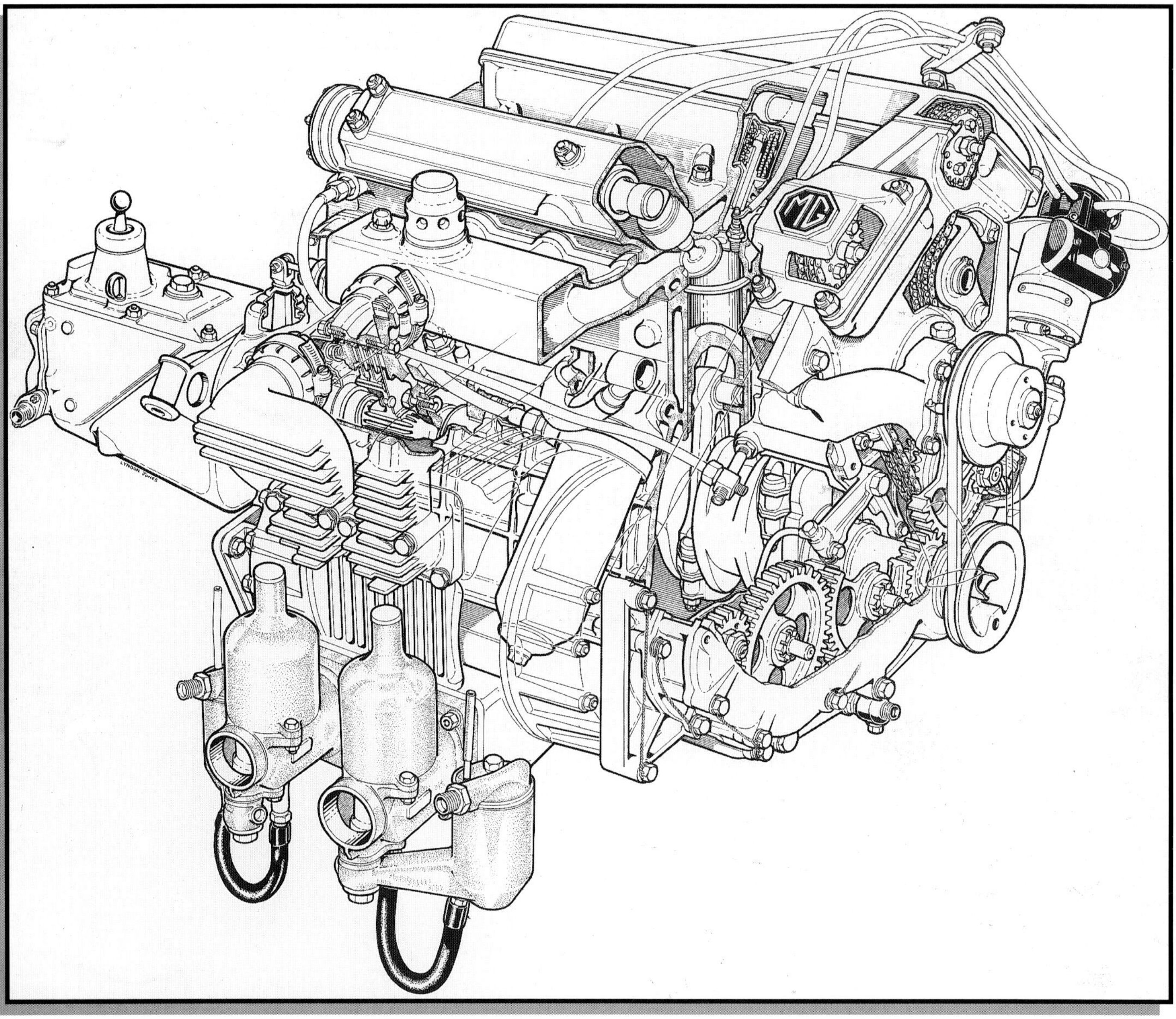

The Shorrock supercharger bolted to the Twin Cam block which took EX 181 to more than 250mph.

Now he had to follow, at more than four miles a minute, a black oil slick laid in the salt.

The car had been exercised on the salt by the back-up driver, Phil Hill, and after days of delay due to rain, Moss finally broke five international Class F records.

1 kilometre	245.64mph
1 mile	245.11mph
5 kilometres	243.08mph
5 miles	235.69mph
10 kilometres	224.7mph

"It all seemed unreal", he said later in his book *My Cars, My Career* "with nearly 300 horsepower hurling me along the blurry black line on otherwise totally featureless salt".

Quite rightly, much was made of the achievement in the British press, with publications like *Autocar* pointing out that it would enhance MGA sales in America.

The team returned in 1959, with the engine bored to 1506cc, to attempt the Class E records. Phil Hill, now driving for Ferrari, was released by his team and took six records including the flying kilometre at 254.91 and the mile at 254.53. The 250mph barrier was finally broken.

However, the triumph was tarnished by an astounding attack on MG's record-breaking activities by Kenny Miles, who had driven for them at Le Mans. In September of 1959, in a US West Coast publication called *Competition Press*, he roundly berated the factory for spending money on break-

The "Roaring Raindrop" on the runway at Brize Norton before its test runs. On the left are Col. E A Loberg, the base commander, Tom Haig and John Thornley.

Tom Haig at speed in EX 181. The driver's vision was very limited.

ing records rather than developing a competition car. He talked of money being frittered away at Bonneville each year which could be spent to much better effect in racing and lead to a better production car. Referring to what was supposed to be Abingdon's competition car, the Twin Cam, he went on: "An attempt to graft a dohc cylinder head on an already highly-stressed push-rod crankcase was obviously bound to lead to trouble, yet under the laboratory conditions of Bonneville the engine functioned perfectly.

"How little similarity exists between [this] and the frantic struggle of a race is well-illustrated by the dismal failure of the dohc MG racing event.

"Even when this engine can be persuaded to stay in one piece, it appears to be incapable of hauling such a large accumulation of Abingdon's sheet metal and cast iron around in company with the admittedly more expensive, but much more reliable, products of Stuttgart". [Porsche].

Coming from someone so closely associated with MG, this tirade was wounding in the extreme to Abingdon and John Thornley quickly replied point-by-point in *Safety Fast!* saying that the amount of money spent in 10 years of record-breaking would not support a racing stable for a single season. But, bearing in mind that the Twin Cam was conceived to compete with the "products of Stuttgart", what he said next was significant: "We never expected the Twin Cam to live with expensive lightweight machinery ... Even a half-wit who compares the weights of the vehicles knows it isn't on before he starts and, if he tries to make it do so, he can expect the engine to protest. The Twin Cam is what it originally set out to be, a darned good Grand Touring car which offers more performance per £ sterling or dollar and gallon of gas than any other machine you can buy and, on that basis, it is good value in anybody's language, even American".

The GT appellation was new in BMC's lexicon. What had happened to the car which, according to publicity at its launch more than a year previously, was "a great British challenge to the world of motor sport" and was going to make "a tremendous impact on competition circles"?

EX 135 (1934-52), EX 179 (1953-57) and EX 181 (1957-59) are all preserved in the UK at the British Motor Industry Heritage Centre at Gaydon in Warwickshire.

Le Mans - the other years

One of the most beautiful MGAs ever built was to have run at Le Mans in 1956 to try and capture the Biennial Cup (which required participation in the race in two successive years) but the project was cancelled because of BMC's withdrawal from racing after the tragedy there.

The car had a standard MGA chassis, but with a de Dion rear end designed by Terry Mitchell and a sleek body drawn by Syd Enever and built by Midland Sheet Metal at Nuneaton. It ran with a twin cam engine and four-wheel disc brakes and had a huge, 21-gallon fuel tank in the rear.

The project was designated EX 186 and the car languished at Abingdon until it was sold in the United States three years later. It is believed to have been bought and retained for many years by a reclusive couple living near San Francisco.

Another twin cam-engined car which was built at the same time, EX 183, had a revolutionary tubular chassis designed by Terry Mitchell and carried a lightweight MGA-shaped body. It's believed to have been broken up.

The next time that an MGA appeared at Le Mans was in 1959; a so-called private entry by the North West Centre of the MG Car Club.

A Twin Cam roadster with a light alloy bodyshell was built and prepared at Abingdon and tested at MIRA by Tom Haig and Henry Stone. The engine was tuned for 130mph and fitted with an undertray to reduce drag. Ted Lund, who had driven at Le Mans in 1955, shared the car with Colin Escott, who hit a dog on the track at 136mph. The accident caused front-end damage which led to severe overheating and eventually the gearbox locked up. That was the end of the 1959 race for SRX 210, but the following year Lund and Escott were back again, with the same car in a different configuration. It appeared as a 1762cc fastback coupé, styled by Don Hayter. It had lower gearing which made it faster around the bends. This time it ran without fault, averaging 91.92mph and finishing 12th overall and first in the 2-litre class.

The car was back for the last time in 1961, this time with a new nose and

The scale model which started the EX 186 Le Mans project. The fairings around the wheels are similar to those used on the record-breaking "Roaring Raindrop", but did not appear in the prototype; neither did the single competition windscreen, which was replaced with a full-width screen. The pictures behind are of works Minis: 33 EJB was Paddy Hopkirk's car in the 1964 Monte Carlo Rally.

The following four unique pictures of the 1956 Le Mans contender were taken inside the Abingdon works by Dennis Williams, who worked in the drawing office. The knock-on Twin Cam wheels are light alloy rather than steel and the exhaust exited at the side. PJB 147 was a press demonstrator.

The streamlined fairing and head-rest. The cover over the passenger seat appears to be canvas rather than metal as in the 1955 Le Mans cars. Rear lights were standard MGA pattern. Note the registration MG1 on the car in front: this was John Thornley's number which was used on several cars, including a green Twin Cam.

The bonnet line was so low that a bulge had to be built in to clear the top of the Twin Cam engine.

The instruments were standard BMC dials except for the over-size Smiths rev counter seen through the MGA steering wheel.

Ted Lund's car, rebodied as a coupé by Morris Bodies, in the pits at Le Mans in 1960. The large driving lamps were from a range sold by his father.

Bob Olthoff instead of Escott. Alas, Olthoff did not get his chance: before the second hour of the race had passed, the Twin Cam had 'put a leg out of bed' as a con rod came smashing through the block.

Thirty years on, SRX 210, always known as the 'Ted Lund MGA', was rebuilt and was racing again at the time of writing.

Tuning the MGA

"The best boost in go for the least in dough" - Road and Track

From the time the MGA first appeared, people tried to make it go faster. Though its performance was more than respectable, there was a small range of tuning kits available - some from BMC itself.

Back in 1961 with a new nose. The car has been restored to this specification.

Lund had planned to run the car at Le Mans again in 1962, but withdrew before the event. The form shows that a non-standard clutch was fitted to the car.

24 HEURES INTERNATIONALES DU MANS

Challenge mondial de Vitesse et d'Endurance

DEMANDE DE PARTICIPATION (4)

Conformément aux articles 1 § 3, 4 § 1 et 40 § 1, la demande de participation ne deviendra **engagement** qu'après décision des organisateurs, décision qui sera notifiée aux concurrents au plus tard le 10 mars 1962.

Je soussigné TED LUND

adresse (complète) PRESTON RD. GARAGE, COPPULL, NR. CHORLEY, LANCS, ENGLAND Téléphone : COPPULL 285

désire participer aux :

24 Heures du Mans 1962

(XXXe Grand Prix d'Endurance)

avec le véhicule
(1) «PROTOTYPE» ou «EXPÉRIMENTAL» conforme à l'annexe A du règlement général des 24 Heures
(1) ~~GRAND TOURISME conforme à l'annexe J du Code Sportif International~~

en vigueur au 1er janvier 1962 et dont ci-dessous les caractéristiques (2) (3)

a) Marque MG Type du véhicule MGA TWIN CAM
Puissance fiscale 108 BHP Année de création 1960 Body 1962 engine
b) Appellation du catalogue MGA TWIN CAM.
c) Nombre de cylindres du moteur 4
d) Marque du moteur MG 1762 c.c. (1) Avant ou Arrière AVANT
e) Cylindrée avec mention de l'alésage et de la course 79.37 mm X 88.9 mm.
f) Carburateur et sa marque WEBER Capacité réservoir carburant 20 GALS IMP
g) Dispositif de suralimentation NIL.
h) Genre d'allumage et sa marque LUCAS COIL
i) Boîte de vitesses (genre et marque) HAUSSERMAN CLUTCH, 4 SPEED MG GEARBOX
j) Freins (genre et indication du genre du servo-freins utilisé) DUNLOP DISC (NON-SERVO)
k) Amortisseurs (marque et nature du fonctionnement) ARMSTRONG PISTON.
l) Roues DUNLOP DISC Marques DUNLOP Indépendantes ou non IND. AV. et AR. RIGID
m) Dimensions des pneumatiques 5.50x15 FR. 6.00x15 R. (marque et genre) DUNLOP
n) Batteries d'accumulateurs (genre et marque) LUCAS 12V
o) Dispositifs d'éclairage avec leurs marques 2 HEADLAMP, 4 TAIL LUCAS.
p) Genre du pare-brise et matière du pare-brise LAMINATED GLASS
q) Noms des conducteurs, titulaires ou suppléants (par ordre alphabétique)
TED LUND MICHAEL REID.

(1) Rayer la mention inutile.
(2) Établir un bulletin par voiture.
(3) Il ne sera tenu compte que des demandes de participation spécifiant la marque, le type et la cylindrée de la voiture.
(4) **Joindre à la demande la liste des conducteurs envisagés et leurs palmarès.**

Tuning in the fifties was commonplace, since the performance of most family cars was dismal. In 1957, 15 different firms were offering go-faster kits for Fords in the UK, and 11 were selling Morris conversions.

The choice of after-market modifications for the MGA 1500 and 1600 from outside suppliers was much smaller, since Abingdon offered a comprehensive range of special tuning products and advice on different levels of tune from stage one to stage five.

Stage I: needed no parts but entailed polishing the ports and grinding the inlet and exhaust manifold ports to match those in the cylinder head.

Polishing the combustion spaces in the head was also recommended and substitution of CC needles in the 1500's carburettors.

The gain was said to be about 3bhp, giving 75bhp in the 1500 engine at 5750rpm and 82bhp in the 1588cc engine. No figure was given for the later engine.

Stage II: for improvements in initial and middle-range acceleration, the camshaft and distributor from the Riley 1.5 were recommended. These were said to give improvements of 2 to 3bhp in the lower ranges, with some attenuation of top-end performance.

Stage IIa: for better mid-range performance, Stage II was followed by the use of flat-topped pistons giving a higher compression ratio - 9:1 for the 1500 and 9.25:1 for the 1600. These

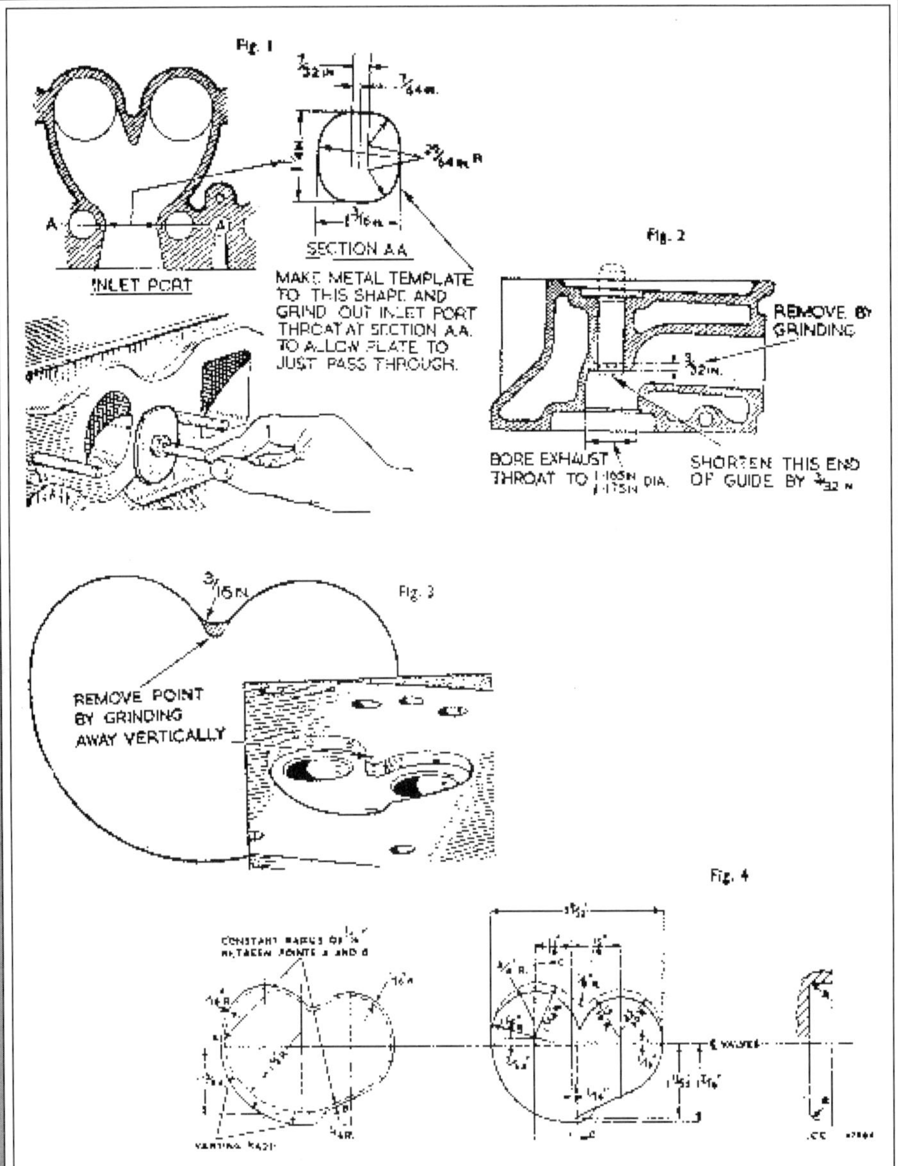

Machining dimensions—enlargement of cylinder head combustion space

Note.—·020 in. (·508 mm.) may be removed from the standard cylinder head face

The basic tuning guide issued by MG suggested opening up ports and combustion chambers. But a high degree of accuracy was required to remove amounts like 3/32nd of an inch at vital points.

pistons required new connecting rods.

Static ignition timing ranged between 4 and 2 degrees before top dead centre for the 1500 and between 3 and 1 degrees for the 1600.

Stage III: action to be taken - port polishing as in Stage I with the changes to compression ratios as in Stage IIa. Static ignition at approximately 4 degrees BTDC and carburettor needle CC in the 1500 and needle number 6 in the 1600. The benefits were said to be an output of 78-80bhp in the 1500 and 85-86bhp in the 1600 at 6000rpm.

Stage IIIa (for the 1600 only): larger carburettors were added to the Stage III tune. These required needles KW (standard), RF (richer) or KW1 (weaker). A new manifold was also needed. There were no air cleaners available to fit the new diameter carbs, so boring out the old ones was suggested, or the use of ram pipes. Output was now 88bhp at 6000rpm.

Stage IV (for the 1500 only): fitting raised top pistons; matching the combustion space needed some grinding away of the area. The pistons had increased clearance over the standard versions and there was a warning that increased oil consumption might occur. Output: 86bhp at 6000bhp.

Stage IVa (for the 1500 only): Stage IV plus the larger carburettors and modifications in Stage IIIa.

Stage V (for the 1600 only): this was the maximum tune for competition purposes which gave 94bhp, without a fan and using 100-octane fuel. It was achieved by tuning to Stage IIIa then grinding out more space around the valve heads, fitting larger valves and skimming up to 0.30ins (0.762mm) off the cylinder head face.

More details of these modifications are given in a British Leyland Special Tuning leaflet C-AKD 819F.

Other special equipment available from Abingdon included competition brake linings, clutch and valve springs. On the standard engine valve bounce occurred at 6000rpm, but springs were available which allowed 6600rpm to be reached, though at some cost to cam wear.

MG produced a range of equipment, not generally available to the public, for those indulging in serious competition, though the choice offered

continued on page 113

Early dashboards were plain painted metal.

The Twin Cam and 1600 Mk II had the facia and scuttle covered in Nuvon and a chrome finishing strip under the dash. This car has had an ammeter added.

The original grille.

The change on the last cars to a recessed pattern for the grille.

The first brochure referred to the MG series MGA, a description not generally used to describe the car.

An early American brochure which featured a car in the rare Tyrolite Green body colour.

The original rear light.

The Mini lamp turned sideways which appeared on the 1600 Mk II.

Larger plinth to accommodate the flasher lens from the 1600.

MGAs were used to sell everything from life insurance to shirts to an American public that considered the car a glamorous 'life style' accessory.

Right: This charming brochure illustration is slightly spoilt by the somewhat childish drawings, which gave the impression of toy cars.

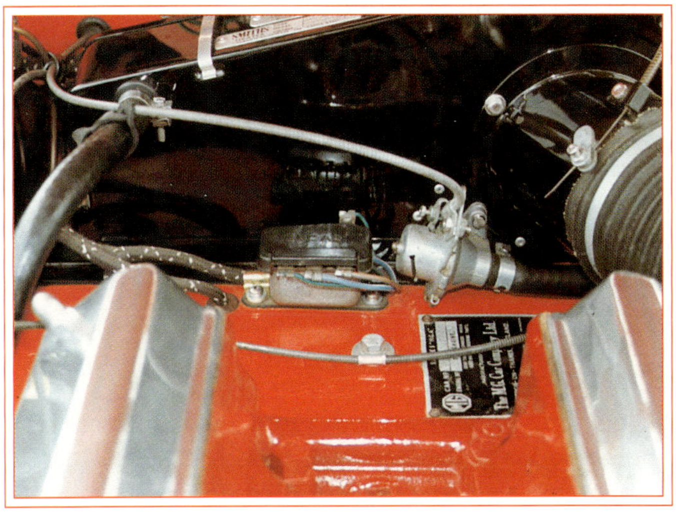

The list of MGA optional extras was long; but few owners specified the adjustable steering column, and even fewer the headlamp flasher, seen here on a Twin Cam.

Many owners have had their car's wire wheels chromed. This car regularly wins concours prizes in the UK.

There are hundreds of events every season for MGA owners and, in Australia and the US, owners drive huge distances in order to take part in club events like gymkanas and concours.

A rare sight: identical cars with consecutive numberplates. The prefix is one allocated to restored/imported cars and was originally used in the Welsh county of Radnorshire.

BMC has the inside track on outdoor fun!
From Sugarbush Valley to Palm Springs . . . everywhere better sports are found . . . you'll find more BMC sports cars than all others combined. And small wonder! All three of these precision engineered beauties are made by BMC, world's largest and most respected sports car craftsmen. Give yourself a lift! Test drive one of these sizzlers today and experience the exuberance of BMC's inside track on outdoor fun. Your BMC dealer will arrange it.
SPRITE
Austin Healey with disc brakes, 2 seats or 4. Removable hardtop available. From $3051.* MG roadster or coupe. Big jet-plane type disc brakes. From $2444.* Austin Healey Sprite, the world's lowest priced true sports car. Removable hardtop available. From $1795.*
*All prices suggested at east coast P.O.E.
MG
SPRITE
AUSTIN HEALEY
MG
The MG Roadster at Vermont's fabulous Sugarbush Valley.
Going abroad? Have a BMC car meet you on arrival. Write for details.
BMC
Products of The British Motor Corporation, Ltd., makers of MG, Austin Healey, Sprite, Morris and Austin cars.
Represented in the United States by Hambro Automotive Corp., Dept. SC-1, 27 West 57th Street, New York 19, N. Y.

Left: With a three-car range priced from $1795 to $3051, BMC hoped to dominate the US market, but met with strong competition from Triumph. The 'jet plane' type disc brakes were a selling point.

The Twin Cam engine (right) was much wider than the standard unit, and made routine servicing very difficult.

761 CJO

Whitewalls may have suited California, but weren't quite the thing for an English country garden!

Whitewall tyres were almost always fitted as standard to US cars, as shown by this American brochure extolling the MGA's virtues.

Below & right: The engine hot air vents on the shroud - one of the MGA's distinguishing marks - were a last minute addition. The original vents were on the side of the car.

ely new — M G A
The smartest,
fastest,
safest
MG
ever
engineered!

1600 Mk II
MGA

Like all car brochure illustrations of the fifties, the lines of the car were rendered with considerable license, making it look much bigger than it was.

Orient Red is the favoured colour for restorations. Few cars survive in the beige and green shades.

A typical British scene in the fifties: an MGA prepares to fill up from hand-operated pumps. This picture was actually taken in the nineties at a museum in Sussex.

Ready for serious stuff ... Mk II coupé in rally trim at the British Motor Heritage Centre at Gaydon.

The first MGA brochure concentrated heavily on the car's engineering and amount of testing done before its launch.

Overheating can be a particular problem in tuned cars. The supplementary Kenlowe fan in front of the radiator is in addition to the normal belt-driven fan.

to everyone was fairly wide. The close-ratio gearbox was often specified as part of the build. The difference in the ratios was:

	Standard	**Close ratio**
1st	3.64	2.45
2nd	2.214	1.65
3rd	1.374	1.268

Rear axle ratio options were:

4.3:1 (Standard)
4.1:1
4.55:1
3.9:1

Other competition options were extra front shock absorbers, (though these could not be used with an anti-roll bar), aluminium oil cooler, racing camshaft and higher-output fuel pump.

Outside manufacturers who sold MGA tuning equipment included Barwell Motors of Chessington, Surrey and Alexander Engineering of Haddenham, Bucks. Both offered cylinder head modifications.

HRG, a well-known engineering firm in Surrey, fitted high compression 9.3:1 pistons and a special light-alloy cylinder head, replacing the siamesed inlet ports - normally on the same side of the engine as the exhaust - with individual ports on the opposite side. This, of course, meant moving the carburettors to the other side as well.

The HRG head was allied to a Derrington exhaust system and *Motor* found in a test run that the converted car was half-way towards the performance of the Twin Cam at much less cost: it also had a 10mph lead on the standard car from 60mph onwards, reaching 100mph whilst the untuned car was still at 90mph.
Performance comparisons were:

	MGA	**HRG MGA**	**Twin Cam**
0-60mph	15.7	13.0	9.1
0-80mph	32.1	23.0	16.2

Fuel consumption - at 28mpg - of the modified car was better than the standard car's 27.6mpg. For some reason the top speed was not recorded.

In the US, *Road and Track's* tune-up man advised that much could be gained for less than $100. He advised skimming the head and removing the air cleaners, replacing them with a made-up steel or aluminium box attached to the carburettors and linked to the existing air intake. The valve clearances were to be 0.002ins less than specified for the inlet and 0.001 wider for the exhaust. These modifications were claimed to give extra urge between 3500 and 4000rpm.

The same magazine also tested a supercharger for the MGA made by Judson of Conshohocken, Pennsylvania, which claimed a power increase of 50 per cent, but the test showed it was about 25 per cent, boosting horsepower from 72bhp to 90 bhp.

The vane-type Judson supercharger gave a manifold pressure of about 5.5psi at maximum and the SU carburettors were replaced with a single Holley. Installation time by the home mechanic of the $260 kit was put at five hours and much was made of the fact that no welding or drilling were required.

Judson's brochure figures claimed 0-60mph in 10.8secs and 0-80mph in 19.6secs.

Road and Track's own comparison figures were:

	Standard MGA	**Judson MGA**
0-60mph	14.2	12.5
0-80mph	29.0	25.0

Top speed was estimated at 105mph and fuel consumption suffered little.

There was also a British supercharger available,which was made by the Performance Equipment Company of Liverpool - Peco, for short - whose managing director rejoiced in the name of Vernon H Farthing and was, it was said, an authority on gas flow. This kit, at just under £100, significantly improved performance in third and top gears by as much as 40 per cent, but petrol consumption in *Motor's* test increased from 27.6mpg for the standard car to 22.8mpg. The blower was of

Judson claimed that their supercharger was silent, easily installed and provided more power at any speed.

the Roots type and belt-driven, with a single SU carburettor attached but, alas, the belt kept coming off so, again, no maximum speed was recorded, though when *Autocar* tried the conversion, it got the MGA up to 102mph.

This supercharger had "slight but not tiresome whine" which was rather more bearable than another blower conversion designed by Sir G Godfrey and Partners. This was said to make a "scissors-grinding noise giving way to an eerie whistle".

None of these proprietary conversions appear to have been taken up with any great enthusiasm - the reason being that, for most people, the MGA was a perfectly satisfactory car as it was.

7

UNUSUAL MGAS

Police cars

Several police forces adopted the MGA as a fast patrol car. The Abingdon design office produced 17 drawings to alter the standard specification to police standards, which included a special accurate Smith's speedometer for catching speeders. To run the extra electrical equipment - such as warning lights and radio - there was a higher output dynamo from BMC's 2-litre C-type engine and a different control box. Larger batteries were used, which required alterations to the chassis where the battery cradles were fitted.

One of the major forces to use the car was Lancashire Constabulary, which had first used MGs in 1932,

One of the first police MGAs, with a November 1955 registration mark, in service with the Lancashire Constabulary.

How did they keep their hats on at speed? The Lancashire force in a newly-delivered batch of 1600 roadsters in June 1959. Note the extra sidelights on top of the wings.

when it bought six J2 Midgets. In those days there were no signs attached to the cars and the hoods had to be kept down even in the winter. When malefactors were about to be apprehended the Midget pulled alongside and an officer held up a cardboard sign saying: "Police - Stop". Lancashire Constabulary took delivery of some of the earliest MGAs produced. The first, XTC 251, a black roadster, chassis 10610, was registered in November 1955. Most of those which followed in the next few years were black roadsters, but there was an occasional coupé and some roadsters in Island Green, Glacier Blue and Orient Red. The later 1600 cars were mostly all white roadsters. One unusual feature on Lancashire Constabulary cars from October 1957 was extra sidelights on top of the front wings, which looked like the torpedo type used on the Midgets.

Lancashire Constabulary bought dozens of MGAs between 1955 and 1960. The car not only gave the force a highly-visible presence on the road but also helped with public relations. A group of some 20 of them used to perform manoeuvres in 'musical rides' at local fêtes.

Britain's largest force, the Metropolitan Police in London, did not use the MGA but, instead, developed a curious attachment to the V8 Daimler SP250, bought mainly to chase motorbike 'cafe racers' around the North and South Circular roads. Scotland Yard did later see the error of its ways and took up the MGB and MGC.

Very odd MGAs

Odd things have happened to some MGAs since they left the factory. A wide variety of engines have replaced the original units; in Cuba, where parts are unavailable, there is a roadster with the engine from a Russian Lada.

Charles Dougherty from Kentucky has taken a 1958 1500 roadster, cut it into six pieces and welded it back together again to make it five inches wider. The car's B-series engine has been replaced by a 5-litre Ford with 5-speed gearbox and the front suspension and steering come from a 1978 Mustang. It's said to do the standing quarter mile in 15 seconds, though even the original could do it in 20.2!

In California, Bob Wing, an enthusiast of electric vehicles who lives at Inverness, north of San Francisco, has converted his 1959 roadster to electric power. His original conversion was powered by golf-cart batteries and an aircraft starter-generator. The latest is a 96-volt DC system using 16 sealed Optima batteries to power a Prestolite 4001 motor. Bob calculates he can achieve a range of 65 miles at 35mph. The car is 500lbs heavier than standard - because of the weight of the batteries. "At least one person a week says 'I had one just like that'" says Bob. "I reply 'not quite like mine' - then, when they insist, I open the hood ..."

Secret MGAs

Although the MGA changed very little during its life, there were prototypes which never saw production, either because they were unsuccessful or because they were being used as a test bed for the MGB.

The 'cheap' MGA

One of the least successful projects was an attempt to build a 'cheap' MG, a return to what the BMC management called a 'buzz box', a no-frills sports car selling at low prices: a modern version of the original MG M-type of the late twenties. Pressure for such a car had been put on BMC by George Bradstock, the influential chairman of MG's London distributor, University Motors. Part of the reason for it was the inroads being made into MG's market by the bubble car, soon to peak with the Suez crisis and petrol shortages of 1956.

Work began on the prototype 'cheap' model as the MGA was launched in the autumn of 1955 and model-making and design work was complete by November. But at Abingdon there was little enthusiasm for the end result, and John Thornley and Sydney Enever realised that they would find it very difficult to build a car down to the price at which BMC wanted to sell it.

It was suggested it be built on the

Where's the familiar B-series engine? Bob Wing's environmentally-friendly roadster has a 65-mile range. but needs batteries in the boot and engine compartment. They are charged from an external supply through, appropriately, the filler cap.

Morris Minor's floorpan, using that model's running gear, but, as Thornley pointed out, unless a great deal of money was spent on retooling the body, the car would end up almost as big as the MGA and be based on an eight-year-old framework.

Other suggestions were for a timber/fabric construction like the thirties cars - that was ruled out because this type of build tended to fall apart in hot climates. There was also an investigation into the use of glassfibre shells - not feasible for the 100 cars a week which were planned.

Earlier, in his development plans for Abingdon dated November 1955, Thornley wrote: "thought has now turned to producing a simple and cheap motor-car based on the MGA. This at first look appears to be a promising line of approach and is being pursued".

The project, codenamed EX 195, used the MGA's body, with a single SU carburettor version of the B-series engine, smaller wheels, the Austin A30 steering box and cheaper trim. It was a car known as the "utility MGA" which no-one at Abingdon liked. But worse was to come. By the autumn of 1956 it was known that BMC were pursuing another 'buzz-box' strategy. "At Abingdon, we are going ahead with the design of an MG based on the 950cc engine" wrote Thornley "primarily because both Longbridge and Healey appear to be doing something of the kind and we have no intention of being left out of the race".

So into the heavy MGA body and chassis went the BMC 948cc A-series engine used in the Austin A35 and Morris Minor. The top speed of this prototype was only about 75mph, a good 20mph slower than the normal production car. The 37bhp engine struggled to propel the MGA. "It was pathetic" said Dennis Williams "and wouldn't pull the skin off a rice pudding. Imagine it, an MG which only went as fast as a baby Austin!"

By March of 1957 the new Longbridge car, the prototype Frogeye Austin-Healey Sprite, using the same engine, was ready. "A brave try" said Thornley caustically and work continued on the MG version. By the spring of 1958 when the Sprite was launched, the MGA version had a Riley 1.5 differential, a Magnette gearbox and 8inch drum front brakes. But performance was dreadful compared with the Sprite, which had twin SUs boosting the A-series to give 45bhp and was much lighter due to its unitary bodywork. There was also little saving to be made in production costs and, given the success of the Frogeye, there was little point continuing. The 'cheap' MGA was dropped.

Also dropped was another unsuccessful 'cheap' experiment of the time at Abingdon, the Riley Wayfarer - a 1.5-litre version of the heavy Riley 2.5-litre Pathfinder. Geoff Allen, who was there at the time, said it was so underpowered it would not run in top gear and ended up as a works hack.

The V4

One of BMC's many experimental engines which never went into production was a V-series of four and six cylinders. These were designed at Longbridge and made in alloy and iron. The V was very narrow - only 18 degrees, making an almost square engine - and the belt-driven camshaft operated across-the-head pushrods as used in the early Bristol engine, a derivative of a pre-war BMW unit.

However, the inspiration for the

The MGA was notorious for its lack of boot space, as this advert illustrates.

engine was not German but Italian. At the research and development department of Longbridge East works, a Lancia had been stripped down, for Lancia knew more than anyone about V engines, having made them since the twenties.

By this time, the BMC research department had been split into two, with conventional engine development in one half under Dr Duncan Stuart and the other half, under Dr John Weaving, experimenting with such bizarre prototypes as a gas turbine-propelled Austin Sheerline. Money was short in both departments.

The V4 was made in two sizes: 1100cc and 2-litre. At least one of these engines was tried in the MGA - acceleration tests were done at MIRA on December 18 1958, but no figures survive. Syd Enever was not keen on the car. "He didn't like the noise it made" said Terry Mitchell "he said the exhaust didn't sound like proper MG".

The engines might have gone into not just the MGA (and the MGB which was then well into the planning stage), but also many other BMC cars; Alec Issigonis considered them for his Austin/Morris 1100 range, but the project was dropped. The investment cost was too great, problems of engine balance arose and there were doubts about the belt-drive technology for the camshaft; although commonplace today, it was then revolutionary. With the A- and B-series engines being reliable, cheap and capable of further development, there was no need for expensive and untried alternatives.

As far as the MGA was concerned, there would have been an additional difficulty. The engine, although of narrow V configuration, was still wider than the standard engine and would not come out through the bonnet aperture if it had to be removed after the car was built ... This meant that the front end of the car would have had to be redesigned.

Other experimental engines were considered at Abingdon. In the summer of 1957 ideas existed for two-stroke engines, a three-cylinder and a V-twin, but these, like many other projects, came to nothing.

Rear suspensions

By 1958, the MGB mock-up was being

Abingdon made an attempt to address complaints about the lack of boot space with this restyling of the rear, seen here in a scale model. The rear of the model, located by dowels, pulls off and can be replaced with the alternative rear end. Fortunately, this design exercise came to nothing.

built and experiments were being done to improve the rather basic rear suspension of the MGA. The Abingdon management had planned to make the MGB a much more sophisticated car mechanically than the MGA, though cost pressures were to ensure that it remained mechanically much the same.

There had been some thought of a de Dion rear end and, indeed, this was built for the Le Mans car EX 186, though it did not run there.

The major experiments for the new car's suspension were tried out on MGAs. These were registered LJB 370 and KMO 326 but, as the works had a somewhat cavalier attitude to number plates, there could have been more than two cars involved.

Work began at the end of April 1958 and was complete by the middle of June. The aim was to get rid of the half-elliptic cart springs at the rear and replace them with coil springs and radius arms to give lower suspension rates and more wheel travel. The designer was Roy Brocklehurst, who came to grief in LJB 370 on March 26 1959, at Boars Hill in Oxford, when a rear wheel was trapped in the arch. The car ended up in the ditch. Brocklehurst was undamaged, the car was not and was rebuilt the following month. At the end of April the set-up was tested at MIRA and also used by Dunlop for tyre adhesion tests at nearby Halfpenny Green airfield.

Various other suspension tuning experiments were tried but, in the end, the MGA's rear end turned up in the MGB, albeit with longer leaves in the springs and softer lever-arm dampers.

The four-seater

Plans to make the MGA in a four-seater as well as two-seater version were being considered for at least a year before the car went into production and the chassis could easily have been adapted. But doubt remains as to how many were actually made - perhaps just the one.

John Thornley thought there would be a good demand for a four-seater, particularly in the United States where no European manufacturer was offering an open four-seater sports car at under £1000. He wrote to the BMC board in August 1954: "In our view, DO 1062 [the MGA project name] lends itself to redesign as a 4-seat car with either a folding head or a rag top. The chassis frame has been designed so as to facilitate elongation amidships and all wings, the trunk unit and the scuttle could be used as they stand."

And if it were not an MG it could be badged as another Abingdon product - a Riley with a different bonnet and grille, but still using the B-series power train.

By the time the two-seater was launched the following year, some preliminary work had been done on a four-seater. Jim O'Neill, MG's chief body draughtsman, remembered that they had done some eighth-scale mod-

Richard Harvey's unique 4-seater - too good to be anything but a factory job?

els to see how it would look - but it was not a success. "You could never have called it a true four-seater as there wasn't that much room in the wheelbase - it was more a two-plus-two".

However, the experimental department appears to have gone ahead with a prototype, codenamed EX 189, the record sheet of which bears the project name MGE (after MGA, the Twin Cam was orginally to have been called the MGB, a large sports coupé the MGC and one of the 948cc 'buzz-boxes' - the Morris Minor version - the MGD). And although no drawings appear to exist, the specification is believed to have included a smaller fuel tank than the Magnette's - it held 9.5 gallons compared with the MGA's 10 gallons, which would have allowed more room for the rear seats.

In 1979, an MGA 1500 came to light in Singapore which could have been this prototype. It was a right-hand drive four-seater, built for export in July 1956. Richard Harvey, who bought it, telephoned the factory at Abingdon, which was still in business, to try to establish its provenance and was told that four or five four-seaters were made for export around 1956, though this has never been officially confirmed. Production records for the chassis number, 17798, do not show any deviation from the standard specification other than the fitting of an adjustable steering column and a rear axle ratio of 4.55:1 rather than 4.3:1. It was despatched from the factory on July 18 and not consigned to the development department.

In the Harvey car, the rear scuttle has been extended backwards to allow for the additional seating, with special brackets to let the hood fold above and behind the back seat. The spare wheel has been relocated, nine inches to the rear, there is a locking bootlid with the fuel filler inside and the rear bumper is two-piece.

The possibility exists that the car was adapted after manufacture by an expert coachbuilder, though experts who have examined it, including Chris Gilbert, a trim foreman at Abingdon, think it was almost certainly done at the time it was built. Richard Harvey appealed in magazines for information about his car and, at the end of 1996, was rewarded with a letter from Peter Knowland who, as an apprentice with Henley's, the Bristol distributor, went on a factory visit to Abingdon in 1956. He remembered seeing in the despatch department several MGAs in different colours in what appeared to be a four-seat or 2+2 configuration. He recalled them because they looked "odd" compared to the other MGAs.

Geoff Barron, of the MG Car Club's MGA Register, thinks that the most significant pointer to Harvey's car being original is the fact that there is no sign of the body ever having a filler neck hole. This assumes that normal production cars had the rear shroud pressed in one operation with the filler hole and hood stud holes being stamped out later. In addition, the aluminium cockpit rails are very well made and have a complex curvature.

Another school of thought is that it is a one-off done by Eric Carter at Morris Motors bodies branch (who designed the coupe), which has a similar profile to the rear scuttle.

Another four-seat option, but not a secret Abingdon project, was a bizarre after-market fitment by the Edinburgh MG distributor, John Gibson. It involved building a twenties-style dickey seat in the boot by rehinging the bootlid at the bottom edge instead of the top. The strengthened lid then supported the back seat cushion. A step was built into the rear bumper to assist access. "A six-foot passenger can stretch out in real comfort", alleged the advertising, failing to mention what effect putting the weight of two passengers behind the rear axle would have on the car's handling. At £67 10s it was an expensive extra which took two weeks to fit and was best avoided.

The four-carb Twin Cam

In June 1958, just before the official launch of the Twin Cam, the development department did some engine work on one of the 1955 Le Mans cars which had subsequently run at Dundrod TT later that year. This was then LBL 301, the car with the droopy nose which had been used to test the Twin Cam unit in competition.

Using one of the factory's many registration numbers - this time LBL 304 - the engine had four carburettors fitted to it, according to notes taken at the time. Rather than being four separate instruments, these may have been two twin-choke Webers, or even the experimental twin-choke SUs mentioned in the Twin Cam chapters.

The headlamp position reverted to normal and the car was extensively tested on the continent, driven to Zurich and Salzburg and run up the hillclimb at Gaisburg. Another MGA, PJB 147, went along as a tender car.

The results of the tests are lost, but it may well be that they formed the basis for fitting Weber carburettors to 1 MTW and 2 MTW, the racing Twin Cams subsequently run by Dick Jacobs, survivor of the Le Mans crash.

The Italian job

In the mid-fifties, BMC became enamoured of Italianate styling, which culminated in a long association with Pininfarina. The first step was taken in 1957 when the Italian coachbuilder, Frua, was commissioned to design and build an MG roadster on the 1500 chassis. It was to be the new MGB rather than a replacement body for the MGA. But when the red car was delivered to Abingdon it was not liked. Though the rear looked neat, the front was garish, with an egg-crate grille, and weight was a problem. Because it had been imported duty-free as a prototype it eventually had to be cut up in front of the local Customs and Excise inspector ...

The rear lights had to incorporate a separate lens for the flasher unit because of changes in British lighting regulations.

8

THE 1600

Time for a change

Nearly 60,000 MGAs were fitted with the 1489cc engine before the first major engine change was introduced in 1959.

The basic B-series engine was now being used in nine of BMC's cars and, over its five years of production, there had been minor improvements to such

When the 1600 was announced in July 1959, the engine delivered 10 per cent more power from an extra 99cc.

things as pistons, rings, oil circulation and sealing, water pump and clutch and gearbox - the type of development expected as experience was gained from the manufacture of hundreds of thousands of engines.

But the change in capacity, which came in May 1959, was unusual in that it was not instituted in any other BMC cars; they had to wait another two years for a larger engine. The only exception until then to the BMC rule of engine rationalisation had been the Twin Cam, and it was the bottom end of this 1588cc unit which was adopted for use with pushrods and a standard cylinder head in June 1959.

The 99cc increase in capacity was achieved by eliminating the water spaces between the first and second pair of cylinders: the bores were 75.39mm rather than 73.025mm while the stroke, of course, remained the same at 88.9mm. *Safety Fast!*, MG's own publicity magazine as it was then, got so confused with these figures that it declared the bore to be 88.9mm - a perfectly square engine! This increase of only 6.65 per cent in displacement gave about 10 per cent more net power and 12 per cent more torque. The peak went up to 79.5bhp at 5600rpm, an increase of 7.5bhp, and torque became 87lb ft at 3800rpm, a increase of some 17 per cent.

Much has been made of the introduction of this new engine concerning 'Appendix J' of the FIA's International Sporting Code, which governed the size of engines used in competition. It has been suggested that it was put into production in 1959 only to meet changes in the code and meet the demands of North American racing. In fact, elimination of the FIA's 1500cc category and its replacement by one of up to 1600cc occurred some three years earlier and, whilst the new engine certainly now qualified under the code, the far more potent 1588cc Twin Cam, which was much more likely to attract competition buyers, had already been in production for more than a year.

The real reason was that, like any car that has been in production for four years unchanged, it was time for a facelift - even though most of the changes were beneath the surface. North American sales of the roadster and coupé combined had dipped from 17,197 in 1957 to 13,235 the following year.

There were several experiments in restyling the front and rear ends. John Thornley admitted later that none of them looked right - they had a "stuck-on" look. It was a tribute to the original design that it could not be improved. What owners needed, Thornley decided, was not different looks but "even-better performance". The most efficient way of achieving this was to use the bigger block from the Twin Cam, even though this engine was unique in the BMC group and required special and more expensive machining on the Longbridge lines.

There were small stylistic changes to the body: a 1600 badge appeared on top of the front wing near the engine compartment ventilators and on the bootlid. To conform with new lighting regulations in the UK which introduced flashing indicator lenses, there were new tail lights on a plinth with separate indicators, and new side and indicator lamps at the front. Cars for export were generally not fitted with the yellow indicator lenses.

On the roadster, the archaic signalling flap was finally done away with and replaced by sliding Perspex sidescreens, the front pane being fixed. A leathercloth envelope behind the seats, provided to store the sidescreens when not in use, was redesigned. There

Stating the obvious; the new sliding sidescreens were a great deal better than the signalling flap, but hardly added "saloon car comfort" as MG claimed. Neither is the hood fit very good in this posed picture.

Below: The classic facia needed no changes.

were small improvements to the hood, and the clutch and brake pedals were repositioned to allow the dipswitch to be moved nearer the clutch pedal.

On the coupé, criticism about lack of space behind the rear seats was answered by pushing the spare wheel backwards so that it was fully in the boot, and cutting back the rear shelf so that a suitcase could be stored there. This, said the BMC literature, "considerably enhanced" the space available. But, naturally, it led to further complaints about lack of space in the boot itself, always a weak point. To try to head these off, an MGA displayed at the Earls Court Motor Show that year sported a Perspex boot panel "showing an unexpectedly large collection of squashy pieces of luggage therein." wrote a *Motor* correspondent hired to give 'the woman's view'. "This is to scotch the old complaint that the car has no room at all and should do something to encourage faint-hearted wives".

The major changes in the 1600 were unseen; it received the more powerful engine and, for the first time, Lockheed 11inch disc brakes on the front wheels and improved linings on the 10inch rear drums. The disc brakes were ecstatically received: "terrific" said *Road and Track*; "superlative" said *Autocar*, whilst *Safety Fast!*, the MG house magazine, described them modestly as: "quite outstanding".

The first 1600, chassis 68851 with blue body/hood and black interior, was completed on June 1 1959 and despatched to North America. The first home market demonstrator, 68861, was finished the same day and two press cars, 69102 and 69103, the following day. Immediately afterwards production of knocked-down kits began for Australia, South Africa and Eire.

MG's London distributor, University Motors, advertised the new model as being available in "gay" new colours - Chariot Red, Iris Blue, Alamo Beige and Dove Grey. These last two are now rarely seen on current MGAs as so many have been repainted in the ubiquitous red.

The magazine testers were quick to note the increased power. *Motor* compared acceleration times with the 1500 and found that 0-50mph took 9.1secs instead of 10.8secs and 0-70mph 17.7secs compared with 21.9.

The magazine was surprised, though, that the bigger engine did not mean an increase in top speed. In fact, in the flying quarter mile, the original 1500 tested had a best speed of 98.4mph and the new car 96.3mph. However, as it pointed out, what mattered was not the ultimate speed, but the ease and rapidity with which the car reached 80mph. Nevertheless, it did get the car up to 100mph in one burst, as did *Autocar,* which touched 101.4mph.

Autocar was surprised by something else; the new engine was noisier and harsher than its predecessor. This may have been due to the test car being less than well-prepared, for the gearbox was very stiff and required two hands to select reverse! "In a thousand miles or so, this would be a pleasant 'box to manipulate" said the magazine mildly. That shows just how much expectations have changed.

Though the car's agility, flexibil-

Just as well the horn push is on the facia ... a 1960 publicity shot which may not have been released since it made the car interior look very cramped.

ity and stopping power were universally praised, some niggling criticisms began to creep into the reviews. *Road & Track* said it was carping rather than grousing by complaining about totally inadequate ventilation, lack of luggage space, poor seats, impractical battery location and lack of a boot handle, synchromesh on first and dual exhaust system. Then it forgave everything by saying that although these annoyances had continued from year to year without correction, the MGA 1600 was an honest automobile built to sell for under $2500, which succeeded because of its value, stamina and durability.

Motor didn't like the flutter in the roof fabric at speed and the complicated nature of the fold-down roof. *Autocar* complained about wind roar and both magazines returned to the topic of lack of boot space.

But, in the final analysis, all would have agreed with *Motor Trend* that the additional power gave extra sparkle without changing the solid character that made the MGA so popular. And said *Motor Sport,* whose editor, Bill Boddy, had an unrivalled ability for summing up the character of a car in a sentence, the car was: "Simple, comparatively outdated, but an eminently enjoyable, very fast sports two-seater".

It was a better car which cost the same as the 1500 it replaced - prices in the UK were £940 7s 6d for the roadster and £1026 15s for the coupé. In the US it sold for $2485.

Few changes were made during the car's two-year life, apart from the introduction of the de Luxe version.

Introduction of the 1600 in 1959 reversed the dip in MGA production and more than 31,000 were built in just under two years until it was superseded by the Mark II. In fact, 1959 saw the peak of MGA production with 23,319 cars coming off the Abingdon lines - and of these, more than 19,500 were to North American specification.

The 1600 Mk II

Sales of the MGA 1600 began to dwindle alarmingly in late 1960. In the year as a whole, nearly 17,000 cars were produced but, the following year, output slumped to a mere 6000 or so. Part of this was because Abingdon production capacity was being taken up by the Morris Minor van and the new MG Midget, launched that year, but it was also due to falling sales, as

Left: The somewhat strained prose of this US advertisement was intended to appeal to the younger driver. Few cliches were left unused ...

Right: ... whilst British audiences got a sober recital of the car's values.

new sports cars, like the Sunbeam Alpine and Lotus Elite, joined the fight for market share.

It is also significant that across the Atlantic when *Motor Trend* pitted the MGA 1600's replacement against the new Midget, it preferred the smaller-engined car: "While it isn't as fast ... we feel that pound-for-pound and inch-for-inch it offers a shade more performance and a whale of a lot more fun. For another thing, it is quite a bit cheaper initially, and much, much more economical to operate. Besides that it comes closest to having that certain indefinable something that made the TC the car that it was". These remarks, particularly about performance, might be considered rather idiosyncratic given that the magazine's own figures showed the MGA was 7.2 seconds faster from 0-60mph and 14mph faster at the maximum. However, the point was well made that you could have a lot of fun for the Midget's $1995 price tag compared with the MGA's $2599.

John Thornley had originally planned a three-year life for the MGA; by 1961 it had doubled that estimate and was becoming outclassed. The delay in replacing it was due largely to a shortage of resources and because the development of its successor, the MGB, took longer than intended.

Plans for a space-frame chassis, V-engine, fuel injection and de Dion rear end for the MGB were all abandoned.

Work on the MGB was due to begin almost as soon as the MGA was launched, though it was not until March 1958 that the first drawings were done. By mid-1960 the first MGB prototype was completed, with a projected launch date of autumn 1962. In the meantime, the MGA had to be kept fresh and, after two years of 1600 production, the stop-gap was launched in June 1961; the clumsily-named but more powerful 1600 Mark II (the 1600 was never officially called the Mk I).

Like the 1588cc car it replaced, an engine size unique to the MGA, it was also not a 1600 but of 1622cc. Designer Eric Bareham re-cored the B-series water jackets once more to make room for larger bores of 76.2mm. Fitting a larger engine was not an MG choice but a BMC corporate move to upgrade its medium-range cars, since the 1622cc unit was to go into all the BMC's Farina cars, the MG Magnette, Austin A60, Morris Oxford, Riley 4/72 and Wolseley 16/60.

The first MGA Mark II chassis was laid down on March 15 1961. The works demonstrator, chassis 100352, was completed at the end of May and licensed on June 14. The car with the following number - 100353 - had been

Left & below: The last of the line, the 1600 Mk II cars pictured at Oxford airport, a few miles from the factory. The only external change was the recessed grille. The more powerful engine was an upgrade for all the BMC mid-range cars rather than a MG initiative. Nuffield always penny-pinched by painting registration plates rather than using the better-looking raised letters and numbers.

completed well before this. It was blue with black trim and hood and was despatched to North America on April 5. The first coupé was completed on July 21.

When the original 1489cc engine was increased to 1588cc to become the MGA 1600, most of the benefit in increased power came from the extra cubic centimetres. With the Mark II, there was again an increase in cubic capacity - by 34ccs - but this time it was other engine changes which provided the major benefit - a 12 per cent increase in power. Compression ratio was raised from 8.3:1 to 8.9:1 (though the higher ratio was an option) and was achieved by using pistons with flat tops rather than the previous slightly concave crowns. There were changes to the head; combustion chamber walls near the valve heads were cut back to increase chamber volume and the inlet and exhaust tracts were improved after gas flow testing. The inlet valves were 0.63ins larger and made of better steel. These improvements lifted maximum output from 83bhp at 5600rpm in the 1600 to 90bhp at 5500rpm.

To cope with the increased power, the connecting rods were stiffened and used larger gudgeon (piston) pins. The crankshaft webs were thickened by and, although this meant a narrowing of the main bearings, it was claimed there was enough reserve in the copper-lead bearing surface area to cope with this.

Other mechanical changes included a flywheel lighter by eight pounds and, more importantly, an increase in the rear axle ratio from 4.3:1 to 4.1:1. This was said to be possible because of the increase in available torque and raised the road speed in top gear from 17.2 to 17.9mph per 1000rpm.

Alterations to the body were very minor, as the production would be running out in a year's time. The one which attracted immediate attention was the front grille, where the angled slats had become vertical, a development not to everyone's taste. Adoption of this recessed grille seems to have been an attempt to make the car look different without spending much money. Another view is that in North America, Detroit's behemoths often backed into parked MGAs, denting the grille slats.

At the rear, more changes in lighting regulations meant the separate direction and stop/tail lights were replaced by an all-in-one unit from the Mini, which was mounted sideways and looked very odd to some.

"Mk II" was added to the 1600 badging and, internally, the facia and

Old Number One was a modified Morris Cowley and the first MG to be built specifically for sporting events. It first appeared in 1925 in the Lands End Trial. Thirty-seven years later, the 100,000th MGA was displayed at the New York Motor Show.

scuttle panel were covered in waterproof Nuvon, similar to the vinyl material used on the coupé. Seat belt attachments were welded to the chassis and pressed into the propeller tunnel, but the belts themselves were yet another optional extra. Other desirable extras were a competition windscreen, anti-roll bar, wire wheels and competition seats. Basic extras were a heater (not standard on MGs until 1968), windscreen washers, adjustable steering column and headlamp flashers. Specifying these added another £26 to the total price.

It was also put about - but not very loudly - that the car was available with Dunlop disc brakes all round - the de Luxe, again.

UK price of the Mark II roadster was £940 7s 6d, the same as the 1600 at its launch two years previously. In five years, the price of the roadster had increased by only £100, though this was not a true indication of factory cost because of changes in purchase tax, which was now levied at a lower rate. In the US the base price before local taxes was $2499. Extras included a heater at $65, wire wheels $100, tonneau cover $35, windshield washer $15 and whitewall tyres $35.

Autocar reported that the new engine was a lively and responsive unit, but not smooth and refined, with distinct roughness at about 5000rpm.

The increased power and torque made only a marginal difference to the MG's performance. Acceleration through the gears showed a reduction of only half a second to reach most speeds up to 70mph, but the time to reach 80mph was cut by two seconds. Despite a wildly-optimistic speedometer, the MGA was again confirmed as a true 100mph car with a maximum of 102.3mph. In addition, the increase in the rear axle ratio allowed a 10mph higher cruising speed of 90mph.

The verdict was that the modifications allowed a little more performance in an effortless manner. "The MGA's traditional character, which appeals to so many, is unchanged" concluded *Autocar*.

In a test done a year before its head-to-head with the Midget, *Motor Trend* found little fault with the Mark II, apart from an unpleasant amount of heat coming though the transmission tunnel whilst being put through its paces in Southern California. It concluded that, as a fun car with basic driving pleasures, it was hard to beat.

After a year of production, in March 1962 there occurred a significant event - production of the 100,000th MGA. It was the greatest number, by far, of any model that MG had produced; the previous highest one-model output was the 29,000 TD Midgets made between 1949 and 1953. The MGA figure was to be subsequently dwarfed by the production of half a million MGBs but, in 1962, the passing of this 100,000 figure - which made it the world's best-selling sports car to date - was considered a significant accomplishment.

As more than 80 per cent of MGA production went to North America, it was fitting that the 100,000th car was displayed there. It was shipped to the US aboard the Queen Mary in March 1962 for the New York Motor Show. It appeared on the stand finished in red, with white trim and whitewall tyres (the US distributor had taken a dislike to the original colour scheme of gold and the car was repainted before it appeared in New York). The air intakes on the front wings bore the usual 1600 Mk II badges with the additional script 100,000 beneath them.

From the 100,000th MGA to the end of production was a short run of 1081 cars. A few weeks later, in May 1962, the first production MGB, an Iris Blue roadster, was built. The last MGA, chassis 109070, finished in the same colour with black trim and a blue hood, went down the line.

The era of MGs with starting handles and a separate chassis had come to an end. The MGA had become the world's best-selling sports car, earned millions of dollars of much-needed foreign exchange and turned a cottage industry in a Thames-side town into a major mass-production facility. Equally importantly, perhaps, Syd Enever's creation has earned the continuing respect and affection of drivers all over the world for its classic lines and for epitomising MG's motto *Safety Fast!*

9

BUYING, RESTORING & RUNNING

Buying

Of all the MGAs made, about 5 per cent were delivered to UK owners and 80 per cent exported to North America. The rest of the world took the remaining 15 per cent. Original registration UK cars are, therefore, rare. Over the past 10 years or so the export drive has been in reverse, with thousands of MGAs coming back from North America to be restored in their home country.

In the boom years of the late 1980s, there were undoubtedly very many which were repatriated as speculative ventures; given a quick cosmetic job and sold on at a profit. Others were restored at some expense in the hope that a concours car would fetch big money as the classic car market continued to climb. Then came the crash, and though MGA prices did not fall as much as those of some of the more exotic marques, there are still roadsters and coupés which have had far more spent on them than they are likely to be worth in the future.

By the mid-1990s, the market had stabilized, with a supply of restored and unrestored cars available at what appeared to be reasonable prices. However, it's the view of MGA expert, Bob West, who imports and restores the cars at Pontefract in Yorkshire, that most of the best MGAs in the dry states of the US were cleared out in the late eighties.

For those thinking of buying direct from abroad rather than through an importer, there are cautionary tales. Cars have arrived at the docks in Britain with engines from Volvos and even Chevrolets. Substitution of MGB components is quite common: not only engines but also wheels, seats and instruments. While the cars will run perfectly well with MGB parts - in fact the engine drops straight in and gives more power than the standard MGA unit - those wanting originality should check chassis and engine numbers (see Appendix 4). The British Motor Industry Heritage Trust at Gaydon, Warwickshire, has all the MGA production records and can check the authenticity of chassis/engine numbers and supply dates of production and despatch.

There are many cars available with paperwork that does not add up - registration documents which show, for instance, that the chassis is a 1500 but the engine a 1600 Mark II. Advertisements for cars often mention "matching numbers"; the chassis and engine numbers are the same as on the original documents. However, authenticity still cannot be guaranteed, since it is easy to replace the plates on which the numbers are stamped.

Cars which have been out in very hot climates may not have much rust, but interior trim may all have to be replaced. Other cars to be avoided are those with glassfibre wings, usually fitted some years ago by an impecunious owner as a cheap alternative to steel. And wire wheels may be chrome bolt-ons rather than painted original knock-ons.

Buying blind may furnish a bargain, but the container at the docks can hold a nasty surprise. It's far better to buy where you can view and check the car before purchase, the

Restoration - a case study

A typical example of a car imported from the US and restored in the UK by its owner is the 1600 roadster belonging to Peter Gamble of Derby.

Chassis 68882, made in the first week of production of the 1600 in May 1959, was despatched to the US on June 2. It ended up in Palo Alto district in California, where it was found in 1990 by UK importer Bob West. The last licence plate showed 1982.

The car had had a hard life; two tyres were down to the canvas and, over the years, the original Dove Grey or white paintwork had been covered with red, light blue and dark blue. The bonnet had been left open for some time, allowing rust to spread, the cylinder head was missing, the pistons seized and the crankcase full of paraffin. But the basic structure was sound and most of the rust superficial.

Over two and a half years, during evenings and weekends, Peter took the car to pieces and rebuilt it. The body needed a new bonnet and boot skins, the rest was de-rusted and the shell resprayed in Iris Blue cellulose. A new right-hand facia panel replaced the left-hand dash panel, which was not original.

The cylinder walls were badly worn, so were bored out to +60thousand and sleeved. The missing B-series head came from a Leyland Sherpa van and was fitted with valves from an MGB head. The crank - which had -10 thousanths wear - was taken down to -20 thousanths.

To convert the disc wheels to knock-on wires, Peter exchanged a spare TD block for the correct axle and got a right-hand steering rack thrown in.

Where the car departs radically from standard is the gearbox, which is a five-speed Ford Sierra unit; alterations had to be made to the bell-housing to make it fit. This alters the figure for top gear performance from 17mph per 1000rpm in the standard car to 20.5mph, and the car will exceed 100mph in fourth or fifth gear. Only the gearlever gives the secret away.

Everything, other than renewing the upholstery, was done by Peter and the car was finally finished in May 1994 at a cost of some £13,000. It was re-registered with a British FF series license number, originally from the Welsh county of Merioneth, but now apportioned to restored cars which need an age-related plate. Since then it has taken part in many shows and rallies and is widely admired.

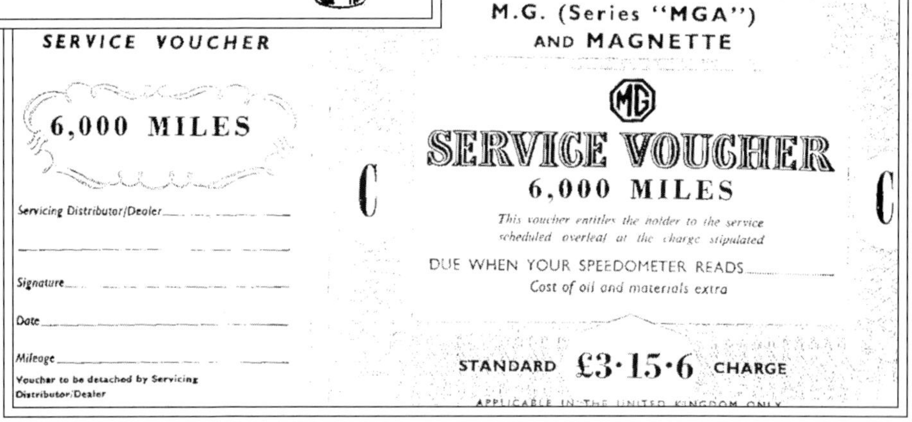

Above & right: The original servicing requirement was every 1000 miles, with major services every 3000 and 6000 miles. Better oils mean less maintenance these days.

Peter Gamble's ex-Californian car as delivered to the UK.

The head was missing and the engine seized.

The dash was not original but would anyway have to be replaced in the conversion with right-hand drive.

The badly-rusted boot floor is typical of restoration projects.

Pristine chassis and engine are mated.

Nearly there ... the front shroud is in place and wings and doors come next.

Finished to concours standard. On the track at Castle Combe in May 1995 during the Norwich Union Classic. (Courtesy John Gaisford)

main criteria for which should be condition and originality.

Restoration - looking for trouble

Although the MGA has a heavy separate chassis, it needs careful examination, particularly at the front, for kinks caused by accident damage. But the model's major enemy is rust: rear crossmembers and battery trays need careful checking. One area particularly noted for decay is in the cockpit, just beneath the sills. Pull back the carpet and check the sidemembers and reach beneath the car and feel the gap between the bottom of the sill and the chassis rail, a small slot where mud builds up, particularly on the side facing the road gutter, depending whether the car is right- or left-hand drive. The damp goes down into the chassis and up into the sill.

Though strictly speaking not part of the chassis, the load-bearing structure in this area - the one-piece sill and pillar assembly, sometimes called the F-section because it looks like that letter on its side - is of particular importance. The doors hang from one upright and close against the other.

The damp caught on top of the chassis rail begins to corrode the sill from within and rust then spreads to the bottom half of the door pillar and catchpost. In bad cases the uprights become virtually detached from the horizontal member. Therefore, doors should be rocked to ensure mountings are sound and careful note paid to how well they close and how accurately they are hung; there should be a one-eighth inch gap all round between the door and the surround. Repairs in this area are expensive and time-consuming because of the difficulty of aligning everything correctly.

Another area of alignment which needs to be checked is the relationship between the car's wings and the shrouds at the front and rear. The shrouds are the panels into which the bonnet and boot fit and to which the wings are attached. The front and rear wings and the shroud are jointed with a grey beading (the same colour on all cars) and damp seeps into the join, rusting the inner wing. Because of the shape of the wings, water runs down the beading, collecting especially at the bottom of the panels. Water also collects on the boot floor and on top of the petrol tank, characteristics shared with the MGB. Rusting in these areas is very common. The floor in the cabin is wood - easily replaced - but floor support rails often suffer terminal rot, remaining fixed to the floor, but not the chassis.

Rust is also found in areas where there are mud traps - below the front headlights and in the rear wings.

Immune from rust should be the bonnet, doors and boot panels, all of which are aluminium skinned. Note that there were two lengths of bonnet - 47 7/8ins and 48 3/8ins. The change was made to accommodate the higher engine on the Twin Cam and the longer bonnet was subsequently fitted to all cars.

The B-series engine is well-known for slogging on despite abuse and many have exceeded 100,000 miles. Signs of age are much the same as with any other engine - smoking exhaust, high oil usage, noisy valves and bottom ends. Normal running oil pressure is about 60psi - worn cam bearings are a major reason for low pressure.

How to tell them apart

Very few readily apparent changes, other than to the lights as shown in the colour section, were made to the MGA during its production life.

There was a change to the radius of the bonnet, after the introduction of the Twin Cam, and this modification affected all MGAs. There are subtle differences inside the front wings after the early 1500s - depressions at the top of each wing to accommodate carburettors and heater hoses, again from the Twin Cam. The 1600 actually had different wings to the 1500 to cope with the change in the front lights.

The Twin Cam was notorious for drinking oil and recently-rebuilt versions still burn oil despite modifications to pistons and rings. This thirst can be abated by fitting modified MGB valve guides with standard MGB rubber stem seals.

In the pushrod engine, cracks in the cylinder head are common, caused either by overheating or improper torque. All model MGA heads are interchangeable, except for the Twin Cam.

Gearboxes wear out their synchromesh but are usually very reliable and easy to maintain. Slipping out of gear, especially on the over-run, is a symptom of selector troubles, which are reasonably easy to cure.

Every part of an MGA can now be supplied by specialist dealers - even the bodyshell. Some parts, like wings, are difficult to fit without considerable adjustment and, due to manufacturing methods, may not be an exact copy of the wing which is being replaced. Trim parts like radiator grilles are often sourced these days in the Far East and quality can be variable. Bumpers, in particular, can be of poor quality and difficult to fit.

Most of the cars imported into the UK from North America have been converted from left- to right-hand drive steering, though there has been a recent tendency to leave the car in LHD form, since it then easier to sell on eventually in continental Europe.

All major conversion parts are easily procured. They are:

Steering rack
Throttle and clutch pedal
Dashboard
Speedometer and rev counter cables
Facia panel

Over the years, many cars have been modified from factory specification, with bodywork changes to accommodate wider wheels and tyres, the fitting of more powerful MGB engines, disc brakes to 1500 models and the use of a Rootes Group remote servo. The trend of late has been to reverse many of these alterations and restore cars to original specification.

Advice to those about to buy an MGA is simple: join one of the main MG clubs beforehand, seek the advice of their experts, study the cars and the market before buying and take counsel from the motor trader - "never buy the first car you see".

MGA

"First of a new line"

APPENDIX 1

SPECIFICATIONS

	Pushrod		
Engine			
	1500	**1600**	**1600 MkII**
Cylinders	4	4	4
Capacity	1489cc	1588cc	1622cc
Bore & stroke	73.025 x 88.9mm	75.39 x 88.9mm	76.2 x 88.9mm
Comp. ratio	8.3:1	8.3:1	8.9:1
Maximum bhp	72 @ 5500rpm*	79.5 @ 5600rpm	90 @ 5500rpm
Maximum torque	77lb.ft @ 3500rpm	87lb ft @ 3800rpm	97lb ft @ 4000rpm
Top gear mph per 1000rpm	17.0**	17.16	17.7
Carburettors	Two SU $1^{1/2}$	"	"
Fuel pump	SU electric	"	"
Fuel tank	10 gallons (45.6 litres)	"	"
Oil filter	Full-flow	"	"
Sump capacity	7 pints (4 litres)	"	"
Water capacity	10 pints (5.67 litres)	"	"
Battery	Lucas SG9E 51 amp/hr 12 volt	"	"
Plugs	Champion NA8	"	"

*(*First cars had 68bhp *at launch)*

Transmission			
Clutch	Borg and Beck A6-G		
Gear ratios:			
Top	4.3	4.3	4.1
3rd	5.908	5.908	5.633
2nd	9.520	9.520	9.007
1st (no synchro)	15.652	15.652	14,924
Reverse	20.468	20.468	19,516
Final drive	Hypoid bevel 4.3:1	Hypoid bevel 4.3:1	4.1:1

Chassis			
Brakes	F&R 10in drum	F. 11in disc R. 10in drum	F. 11in disc R. 10in drum
Suspension			
Front	Coil /wishbone/lever dampers		
Rear	Semi-elliptic leaf/lever dampers		
Steering	Cam Gears rack & pinion		
Wheels	Disc or wire		
Tyres	5.60 -15 tubed		
Equipment			
Intruments	Speedometer with trip recorder, rev counter, oil pressure gauge, fuel gauge, water thermometer		
Dimensions			
Wheelbase	7ft 10in		
Track:			
Front	3ft 11.5in		
Rear	4ft 0.75in		
Length	13ft		
Width	4ft 10in		
Height	4ft 2in		
Ground clearance	6in		
Turning circle	28ft		
Kerb weight:			
Roadster	2015lbs*		
Coupé	2105lbs*		

**dependent on equipment*

Prices at launch			
Roadster	£844	£940	£940
Coupé	£1087	£1027	£1047

Twin Cam
Engine

Cylinders	4
Capacity	1588cc
Bore & stroke	2.96in x 3.5in (75.4mm x 88.9mm)
Compression ratio	9.9:1
Maximum bhp	108 @ 6700

Maximum torque	105lb.ft @ 4500rpm
Max BMEP	163lb/sq in @4500rpm
Maximum rpm	7600
Top gear mph per 1000rpm	17.3
Carburettor	Two SU HD6
Fuel pump	SU electric
Fuel tank	10 gallons (45.6 litres)
Oil filter	Tecalmit full-flow
Sump capacity	12 pints (5.7 litres)
Water capacity	13.5 pints (7.6 litres)
Battery	Lucas SG9E 51 amp/hr 12 volt
Plugs	Champion N4. (N3 high speed)

Transmission

Clutch	Borg and Beck 8in
Gear ratios:	
Top	4.3
3rd	5.908
2nd	9.520
1st (no synchro)	15.652
Reverse	20.468
Final drive	Hypoid bevel 4.3:1

Chassis

Brakes	Dunlop disc; friction lining area 32 sq in (206.4 sq cms)
Suspension:	
Front	Coil springs and wishbone, Armstrong lever dampers
Rear	Semi-elliptic leaf springs, Armstrong lever dampers
Steering	Cam Gears rack and pinion
Steering wheel	16.5in (41.9cms) diameter four-spoke, 2.75 turns lock/lock
Wheels	Dunlop centre-lock steel disc
Tyres	Dunlop Road Speed 4, 5.90 -15 tubed

Equipment

Instruments	Speedometer with trip recorder, rev counter, oil pressure gauge, fuel gauge, water

	thermometer
Tool kit	Tyre pump, wheel nut hammer, tool roll with 2 screwdrivers, 3 spanners, 4 box spanners, adjustable spanner, tommy bar, feeler gauge, grease gun, pliers.

Dimensions

Wheelbase	7ft 10in
Track:	
Front	3ft 11.9in
Rear	4ft 0.87in
Length	13ft
Width	4ft 10in
Height	4ft 2in
Ground clearance	6in
Turning circle	31ft 4in
Kerb weight	2156lbs

Optional extras

Heater	Smiths 3.5kw fresh air type
Radio	Radiomobile

Ashtray, cigar lighter, adjustable steering column, luggage grid, wing mirror, tonneau cover, radiator blind, foglamp, competition windscreen, windscreen washer, hardtop, sliding sidescreens, badge bar, sun vizor, woodrim steering wheel, twin high-note horns,oil cooler, de luxe seats.

Prices at launch

Roadster	£1265 17s
Coupé	£1357 7s

SERIES "MGA"

SERIES "MGA"

LUBRICATION CHART

EXPLANATION OF SYMBOLS

CASTROL XL. A high quality motor oil recommended for the M.G. engine and gearbox in summer and winter.	CASTROL HYPOY GEAR OIL. A powerful extreme-pressure gear oil essential for the lubrication of Hypoid rear axles. Also recommended for the steering gearbox.	CASTROLEASE LM GREASE. A lithium base grease recommended for all grease nipples.

If oil consumption presents a problem after your car has covered a considerable mileage, then it is advisable to use the next heavier Castrol grade in summer.

WARNING. Do not ask for "XL" when you require CASTROL XL. Ask for "CASTROL XL" and see it drawn from a CASTROL container. CASTROL is marketed as a finished product and contains carefully balanced additives in precise quantities. The addition of other additives or diluents cannot improve it and in some cases can prove harmful. While CASTROL grades (except CASTROL R) will mix with other mineral oils, full benefit cannot be derived unless the correct CASTROL grade is used by itself.

Every 1,000 Miles

STEERING TIE-ROD — *1 nipple*
SWIVEL PIN — *2 nipples*
Apply **CASTROLEASE LM GREASE** with 3 or 4 strokes of the grease gun at the nipples indicated.
Note: If CASTROLEASE LM is not available, CASTROLEASE MEDIUM may be used as an alternative.

Every 6,000 Miles

FRONT WHEEL HUB. Remove wheel hub disc and grease cap, fill cap with **CASTROLEASE LM GREASE** and replace.
Note: If CASTROLEASE LM is not available, CASTROLEASE HEAVY may be used as an alternative.

Every 250 Miles

ENGINE. Check oil level and replenish, if necessary, to the **"MAXIMUM"** mark on the dipstick with **CASTROL XL.**
AFTER THE FIRST 500 MILES AND THEREAFTER EVERY 3,000 MILES, REMOVE DRAIN PLUG, DRAIN OFF OLD OIL WHILE WARM, REPLACE PLUG AND REFILL WITH FRESH CASTROL XL.

OVERSEAS
Air temp. consistently over 32° F. CASTROL XL.
Air temp. consistently between 0° F. and 32° F. CASTROLITE.
Air temp. consistently below 0° F. CASTROL Z.

Capacity: 7 pints.

Every 1,000 Miles

REAR AXLE. Check oil level with dipstick and replenish, if necessary, to level of filler plug hole with **CASTROL HYPOY GEAR OIL.**
AFTER THE FIRST 500 MILES AND THEREAFTER EVERY 6,000 MILES, DRAIN OFF OLD OIL WHILE WARM, REPLACE PLUG AND REFILL WITH FRESH CASTROL HYPOY GEAR OIL.

OVERSEAS
Air temperature consistently below 10° F. CASTROL HYPOY LIGHT GEAR OIL should be used.

Capacity: 2¾ pints.

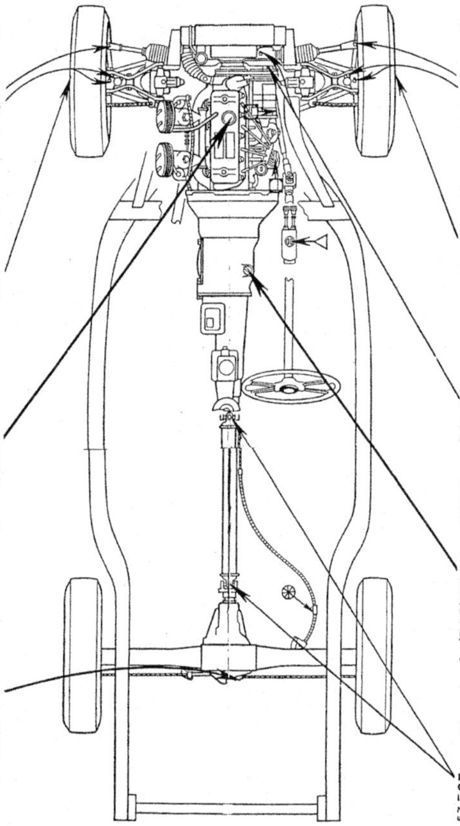

Every 1,000 Miles

STEERING TIE-ROD — *1 nipple*
SWIVEL PIN — *2 nipples*
Apply **CASTROLEASE LM GREASE** with 3 or 4 strokes of the grease gun at the nipples indicated.
Note: If CASTROLEASE LM is not available, CASTROLEASE MEDIUM may be used as an alternative.

Every 6,000 Miles

FRONT WHEEL HUB. Remove wheel hub disc and grease cap, fill cap with **CASTROLEASE LM GREASE** and replace.
Note: If CASTROLEASE LM is not available, CASTROLEASE HEAVY may be used as an alternative.

Every 12,000 Miles

STEERING GEARBOX. 10 strokes only and the PINION SHAFT 2 strokes only with an oil gun filled with **CASTROL HYPOY GEAR OIL.**
Temperatures below 10° F. CASTROL HYPOY LIGHT GEAR OIL.

Every 1,000 Miles

GEARBOX. Check oil level and replenish, if necessary, to the **"MAXIMUM"** mark on the dipstick with **CASTROL XL.**
AFTER THE FIRST 500 MILES AND THEREAFTER EVERY 6,000 MILES, REMOVE DRAIN PLUG, DRAIN OFF OLD OIL WHILE WARM, REPLACE DRAIN PLUG, AND REFILL WITH CASTROL XL IN ALL TEMPERATURES.

Capacity: 4 pints.

Every 1,000 Miles

PROPELLER SHAFT. Apply **CASTROLEASE LM GREASE** with 3 or 4 strokes of the grease gun at the nipples indicated.
Note: If CASTROLEASE LM is not available, CASTROLEASE MEDIUM may be used as an alternative.

ALSO REQUIRING ATTENTION

OIL CAN

Every 3,000 miles. DISTRIBUTOR. *Indicated thus* ▢→ Lift rotor arm off the top of the spindle and apply a few drops of CASTROLITE to the cam bearing. Lightly smear the cam and contact pivot with CASTROLEASE LM GREASE. Add a few drops of CASTROLITE through the hole in the contact breaker base through which the cam passes, to lubricate the automatic advance mechanism.

DYNAMO. *Indicated thus* ▢→ Add 2 drops of CASTROLITE to oil hole in rear end bearing plate. Do not over-oil.

DOOR HINGES, BONNET LOCK AND OPERATING MECHANISM. Apply a few drops of CASTROLITE.

Every 1,000 miles. CARBURETTER. Remove cap from top of each carburetter suction chamber and add a teaspoonful of CASTROLITE.

BRAKE MASTER CYLINDER. *Indicated thus* ▷→ Check level and replenish if necessary with the correct fluid.

HAND BRAKE CABLE. *Indicated thus* ⊛→ Give 3 or 4 strokes with the grease gun filled with CASTROLEASE LM GREASE.

Every 3,000 miles. AIR CLEANER. Remove, clean and re-oil with CASTROLITE. Allow to drain before reassembling.

Every 6,000 miles. EXTERNAL OIL FILTER. Remove, wash out filter casing and fit new element.

HYDRAULIC DAMPERS. Check level and replenish if necessary with correct fluid.

UPPER CYLINDER LUBRICATION. In order to lubricate more effectively the pistons, piston rings, valve stems and guides, WAKEFIELD CASTROLLO may be used in the proportion of 1 oz. to 2 gallons of fuel.

TYRE PRESSURES	
NORMAL	Front 17 lb./sq. in. Rear 20 lb./sq. in.
FAST MOTORING	Front 18 lb./sq. in. Rear 23 lb./sq. in.

This chart has been checked and approved by the Manufacturers:

THE M.G. CAR CO. LTD., ABINGDON-ON-THAMES, BERKS.

Issued by C. C. WAKEFIELD & CO. LTD., GROSVENOR STREET, LONDON, W.1.

This chart shows lubrication and oil change periods when operating under ideal conditions. For less favourable conditions, especially overseas, more frequent lubrication and oil changes are necessary.

APPENDIX 2

COLOUR SCHEMES

This information is taken directly from the colour charts issued by BMC.

Early 1500 and Twin Cam

Body	*Hood*	*Leather*
Orient Red (two-seater and coupé)	Black	Red or Black
Old English White (two-seater and coupé)	Black	Red or Black
Glacier Blue (two-seater only)	Ice Blue	Grey or Black
Tyrolite Green (two-seater only)	Ice Blue	Grey or Black
Island Green (coupé only)	-	Grey or Black
Mineral Blue (coupé only)	-	Grey or Black
Black* (two-seater)	Ice Blue or Black	Red or Green
Black (coupé)	-	Red or Green

**Some early black cars had dashboards in red or green rather than in the body colour*

Later 1500 and Twin Cam

Body	*Hood*	*Leather*
Orient Red (two-seater and coupé)	Black	Red or Black
Old English White (two-seater and coupé)	Black	Red or Black
Glacier Blue (two-seater only)	Ice Blue	Grey or Black

Ash Green (two-seater only)	Ice Blue	Grey or Black
Mineral Blue (coupé only)	-	Grey or Black
Black (two-seater)	Ice Blue or Black	Red or Green
Black (coupé)	-	Red or Green

1600, 1600MkII and later Twin Cam

Body	***Hood***	***Leather***
Chariot Red (two-seater and coupé)	Beige Beige Grey	Red Beige Black
Iris Blue (two-seater and coupé)	Blue	Black
Alamo Beige (two-seater and coupé)	Beige	Red
Dove Grey (two-seater and coupé)	Grey	Red
Old English White (two-seater and coupé)	Grey	Red or Black
Black (two-seater)	Grey	Red or Beige
Black (coupé)	-	Red or Beige

Colours were also available to special order. CKD bodies usually left the factory in primer and were painted abroad.

APPENDIX 3

CHASSIS & ENGINE NUMBERING

Chassis numbering

1500 cars

HDA 23 24609 — Chassis number

Model	**Body type**	**Colour**	**Market**	**Paint**
MGA	D=roadster	A=Black	1 RHD home	1 Synthetic
	M=coupé	C=Red	2 RHD export	2 Synthetic
		D=Blue	3 LHD export	3 Cellulose
		E=Green	4 LHD N.America	4 Metallic
		K=Lt Red	5 RHD CKD export	5 Primer
		L=Lt Blue	6 LHD CKD export	6 Cell/Syn
		R=White		
		T=Lt Green		
		H=Primer		

1600 cars

GHNL 2 10001 — Chassis number

Make	**Engine size**	**Body**	**Steering**	**Version**
MG	1400 to 1999cc (i.e. BMC mid-range engines)	N=roadster D=coupé	lhd (blank for rhd)	variant 2=MkII

Twin Cam

YD1 2611 — Chassis number

Model	Body	Market
Twin Cam	D=roadster	1 RHD home
	M=coupé	2 RHD export
		3 LHD export
		5 RHD CKD (S Africa)

The original numbering system was changed for 1600 cars to the BMC corporate numbering scheme. To add to the confusion, Twin Cams had a separate scheme. Colours were not specified in the chassis numbers after the 1500. Some colours - such as Orient Red - were specified as both C and K.

Engine numbering

As the B-series engine was used so widely, it had a bewildering set of serial and parts numbers.

Serial numbers for the different BMC models were defined by G for MG, M for Morris, W for Wolseley, etc., followed by the first two numbers of the cubic capacity: e.g. 15 or 16 for MGAs. Serial numbers began at 101 at engine change points.

Early MGAs also had an additional engine prefix of BP.

15GB-U-H 37529 — Engine number

Size	Make	Model	Gearchange	Compression
15=1500cc	MG	B 1500/Twin Cam	U=Centre	High
16=1600cc		D 1500 revised*	Da=close	(standard cars
(inc. Twin		A 1600	ratio box	were low, but not
Cam)		C 1600 MkII		marked so)

** From January 1959. Different starter position and modified gearbox.*

Pushrod engine sizes can be even more easily identified by the number cast into the crankcase at the front in large figures: e.g. 1500, 1600 and 1622. Another legend nearby spells out MOWOG (**MO**rris/**WO**lseley/**MG**).

Serial numbers

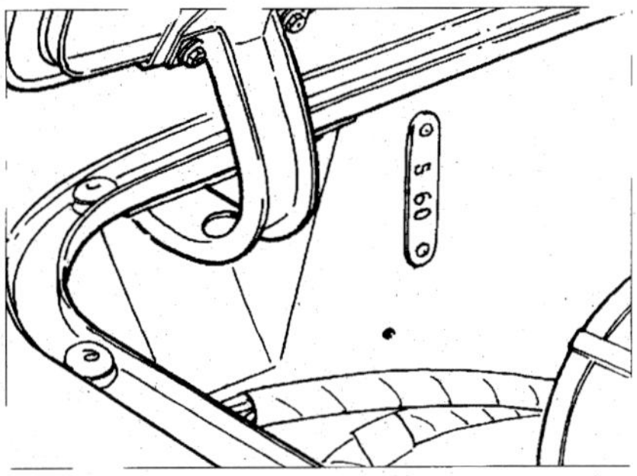

Body number. Stamped on a plate secured to the right hand side of the dash panel

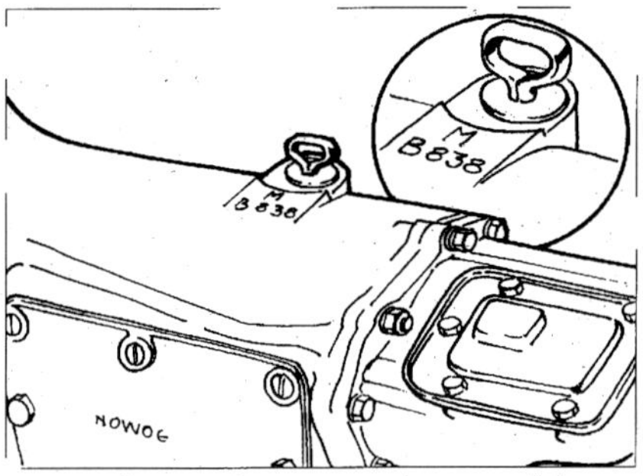

Gearbox number. Stamped on top of the gearbox casing adjacent to the dipstick

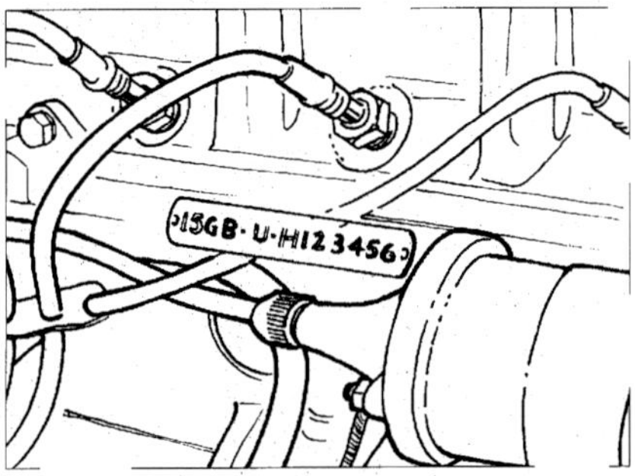

Engine number. This is stamped on a plate on the right hand side of the cylinder block

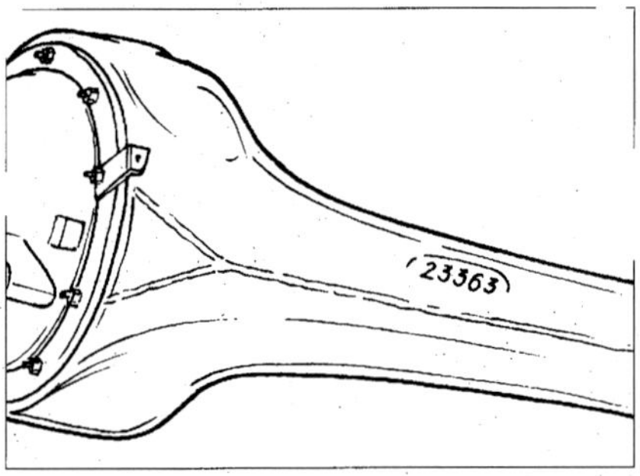

Rear axle number. Stamped on the front of the lefthand rear axle tube

Engine Serial Numbers

Sample

BP15GB/U/H1234

1st	2nd	3rd	4th	5th	6th	
BP	15	G	B	U	H	1234

1st	BP = Prefix for early 1500 only
2nd	15 = 1500 cc 16 = 1600 cc
3rd	G = MG A = Austin M = Morris W = Wolseley
4th	B = 1500 A = 1600 C = 1622
5th	U = Centre gearchange
6th	H = High compression L = Low compression

Courtesy Anglo Parts B.V, Belgium

APPENDIX 4

EX CARS

The EX numbers used by the factory for experimental projects such as race, record and prototype cars can confuse mightily, though some *aficionados* can reel them off.

Here are the main numbers relating to the MGA; they are not chronological.

EX 172 - the George Phillips/Enever-designed special-bodied TD for Le Mans.

EX 175 - registered HMO 6, the rejected prototype MGA with the new chassis.

EX 179 - Bonneville record-breaker with spare MGA chassis and TF engine.

EX 181 - "Roaring Raindrop" with blown twin cam engine. Moss record car.

EX 182 - 1955 Le Mans cars, MGA prototypes.

EX 183 - tubular chassis for 1956 Le Mans car with MGA lightweight body.

EX 186 - 1956 twin cam Le Mans car with special streamline body.

EX 187 - Twin Cam with disc brakes.

EX 188 - four-seater version of MGA, described in register as MGE.

EX 191 - production MGA development - from May 1955 to Feb.1961.

EX 194 - Police modifications.

EX 195 - cheap MGA.

EX 196 - MGA hardtop.

EX 197 - MGA coupé.

EX 212 - Twin Cam coupé.

EX 214 - Frua replacement for MGA.

EX 216 - V4 engine.

APPENDIX 5

CONCOURS MGA SPECIFICATIONS

Many MGA owners go to enormous trouble to make their cars as close to the original specification as possible. Concours judging tends to be an art rather than a science because of variations in standards and specifications, and many new owners are confused about the original appearance of their cars.

These Australian concours judging notes show how MGA originality is established.

Originality

A. Exterior

No points should be deducted for car colour unless it is not period.
Lower sill finisher strip should be fitted from car 19,949 and painted body colour.
Original bumpers were curved and not flat.
Correct type of vent for car model.
Correct indentation on rear beaver below bumper.
Original bumpers were painted silver-grey on the inside.
Check appendix for correct car colours.
Ensure correct items are chromed.
1500s do not have reinforcing on air vents.
Bumper bar ends were suitably curved not straight.
Rubbers should be fitted behind overriders.
Mudguard and grille piping should be grey.
Grille piping should terminate underneath grille in line with each of the 2 small chrome bolts.
Tail light plinths: 1500 black; 1600 grey.
1500 side curtains should be flap-type until October 1956, when sliding side curtains were first offered as an option.
Flaps should have a chromed, spring-loaded damper.
Sliding side curtain frames were vinyl covered. Aluminium type was available with hard tops. If a car has steel side curtain frames painted black it should not be penalised as these were factory replacements.
Hood fabric was vinyl, normally black, but offered also in ice-blue, beige & grey on UK-manufactured cars.
Glass should not be tinted.
Wire wheels should be on all Australian-plated cars. They were offered as an option on English cars.
Chrome-spoked wheels were not a standard option.
Tyres may be cross-plys or radials.
Paint colour on wire and steel wheels should be silver-grey metallic.
1500s pre-48,730 had 'MG' on knock-ons.
Headlights to car no. 58,918 should be Lucas with replaceable globe.
Headlamps after car no. 58918 are sealed beams

Fog lights were optional on all models.
Radio aerial, if fitted, should be retractable into mudguard.
Number plate backing plates to be fitted front and rear, painted black.

B. Interior

Seal must be fitted between wood and battery cover.
Auster plate on inside of windscreen.
Australian-produced cars should be all vinyl.
UK-manufactured cars should have leather on all wearing parts and rails on top of doors; vinyl on all edges and back of squabs.
Carpet should be Westminster fluted-type or Crossley Karvel tufted-type, with felt carpet underlay under front tunnel. 6 pieces in all.
Binding should be on centre tunnel, oil flap and around crossmember only.
Heel mat fitted to drivers side only, should be sewn in and of criss-cross design.
Battery cover should be painted car colour.
Floorboard screws should have cup washers.
Hood bows should be grey or beige in colour with overlay on each bow. Also black was used on factory replacement bows.
Centre clip on all models except 1500.
Boot should not be carpeted.
Spare wheel cover should be vinyl except in UK cars where carpet has sometimes been found.
Turn lever should be black and shorter than T-series.
Knob lettering should be white.
Radio blanking plate should be body colour, except for vinyl-covered dashes.
Vinyl-covered dashes should have chrome around speaker panel and chrome cover strip surrounding bottom of dash.
Grommet fitted where door pull passes through trim.
Piping should be above and below crash pad on top of dash.
Radio, if fitted, should be early transistor-type with separate speaker. Two brands: Radiomobile or HMV for UK-manufactured cars.
Rear vision mirror should be gold colour.
Correct chrome spring to steering column.
1600s onwards had adjustable columns. 1500s were unadjustable with adjustable columns as an optional extra.
Door hinges to be body colour.
Door catches should be zinc-plated, including catch on door itself.
Furflex door frames can be black, or green, grey or red on UK cars.
Pop rivets should hold outer sill to front door pillar. No continuous welds.

C. Engine Bay

Correct oil filter.
Interior should be the same colour as body.
There should be no evidence of underbody sealer or stoneguard, as dealers would not have sprayed engine bay.
The following parts should be painted Carnation Red: engine, nuts and bolts, cylinder head, rocker cover, heat shield, inlet manifold, dip stick, vacuum pipe, top hose mount and water pump banjo.
Coil should be silver with Lucas decal.
Exhaust manifold should be unpainted. Painted black is acceptable.
Fan - can be black, engine colour, or yellow, depending on model (don't penalise for yellow as this was done for safety reasons).
Heater pipes should be natural copper finish.
Heater or heater blanking plate should be black.
Oil filler cap should be original plated type, not MGB or chrome.
Pedal cover and heater cover should be black.
Pedal cover surround plate should face up, not down.
Braiding should be black with white slashes, cotton covered with individual wires.
Colour-coded.
Correct metal clips fitted.
Replacement harnesses and late 1600 Australia-only cars had plastic covered harnesses.
Speedo cables grey, not black.
Yellow tracers on push-rod cars and white on Twin Cams were also used.
Air intake pipe in front of radiator.
If heater is not fitted, blanking plate should be installed.
Heater pipes should be natural copper finish.
Fuel pipes should be braided steel.
All catches should be painted body colour, including pull-rods.
Bonnet support rod should be black.
Cylinder heads: '15' on 1500 and 1600 cars, and '16' on MK IIs (not '18').
Brass plate with firing order on intake manifold
Spark plug leads should be numbered with yellow rings.
Spark plug clips should be correct suppresser-type.
All guard bolts protruding into engine bay should be bright zinc and have a pointed end.
Correct I.D. plates on fire wall.
Patent plates on rocker cover.
One Lucas decal on coil.
Control box 1500s should be silver colour (not chrome-plated).

Hose clips to be two-wire vintage-type.
Bonnet should have wooden stiffener with anti-drum felt behind it
Felt strip on bonnet to meet radiator.
All carburettors had brass tops to dampers.
Body number should prefix with 'B' on 1600s onwards.

D. Underbody

Body spray (stone guard or sound deadening) should only be evident under boot and transmission tunnel.
Floorboards should be painted black.
Screws holding floorboards down should have pointed ends.
Bolts used on car should be plated.
Correct clips should hold loom.
Correct spacer should be used on handbrake clips.
Rear axle should be black.
Gearbox should be aluminium (do not deduct points if differential is painted black as there is a discrepancy between Australian and UK cars).
HP fuel pumps should be rubber-covered up to car no. 93,225.
Correct battery terminals.
Correct steel cable cover over link battery cable.
U-bolts should have extra locking nuts.
Rubber pads must be fitted under batteries.
Exhaust pipe and silencer should be aluminium or light grey or black.
Springs and metal parts should be painted black, including brake callipers and drums.
Shock absorbers can be natural aluminium with black painted arms.
Replacement shocks were green and black, so no penalty for these combinations.
Anti-sway bar was optional from car 66574.
Short and long strap screwed on where gearbox meets engine.
Correct long nuts holding flange to exhaust manifold.
Tie rod ends and rack boots must have proper ring clips.
Lucas high-low type horn/
One horn normally fitted behind front valance/
Sometimes two horns were fitted as an optional extra/
Nuts and bolts should be stamped with the following brands: Linread, Wiley, Newton, Rubery Owen, GRS, etc, NOT AJAX, and should be plated.
Splash guards had split rivets holding rubbers.
Front valance should have rectangular plates behind fixing screws.
Rear shock absorber bolts should face outwards.
Diff and gearbox should have breathers fitted to top.

E. Boot

Tool roll should be strapped to top of spare wheel.
Tool kit complete (there is no information to tell us what was in an Australian tool kit e.g. no grease gun).
Boot pull-felt should be fitted under rod.
Two felt pads should be fitted under spare wheel.
Pull rods should be painted body colour.
Boot lid to have wooden stiffener with anti-drum felt fitted.
Felt strip fitted in gap between boot floor and body.
Support rod should be black.
1600 Mk IIs can have carpet mats in boot.
Tool kit should be hessian-type material. Soft hammer to be on display.
Spare wheel cover grey “Hardura”.
Australian 1600 Mk II had grey “Hardura” as a boot mat.

The notes were kindly provided by Henry Stenning of the MG Car Club of New South Wales but would apply in most instances to MGAs produced in Britain and elsewhere.

APPENDIX 6

PRODUCTION

MGA production by year

	1500	***Twin-cam***	***1600***	***1600 Mk II***	***Total***
1955	1003				1003
1956	13410				13410
1957	20571				20571
1958	16122	541			16663
1959	7644	1519	14156		23319
1960		51	16930		16981
1961			415	5670	6085
1962				3049	3049
TOTAL	**58,750**	**2111**	**31501**	**8719**	**101,081**

MGA model timetable

Model	**First chassis**	**Date**
1500	10101	Sept 1955
1500 coupé	20671	Sept 1956
Twin-Cam	501*	July 1958 (ended May 1960 at 2611)
1600	68851	June 1959
de Luxe	91240	June 1960
1600 Mk II	100352	April 1961
Last MGA	109070	May 1962

**Twin-Cam chassis numbers were separate from the general run of MGA numbers.*

APPENDIX 7

CLUBS, SPECIALISTS & BIBLIOGRAPHY

Clubs

Two major clubs cater for the MG owner and both are based in the UK, though with branches abroad. The clubs welcome all MGs - regardless of age - from the first Midget to the latest MGF, and are a fount of information on spares and technical details. Both run events such as racing, rallying and touring, and have busy social calenders.

The MG Car Club, whose headquarters is now in the original buildings at the defunct Abingdon factory, has a long association with the marque. The club was founded in 1930, run for a time by John Thornley, and was also closely associated with the factory, at least until the British Leyland era. Its monthly magazine, *Safety Fast!*, is dedicated to those who 'practise driving as an art'.

The club is international and subdivided into home and overseas branches. There are several in continental Europe, 6 in South Africa, 14 in Australia and nearly 40 in the US.

In addition, the MG Car Club caters specifically for the MGA owner, with a Register for the MGA and another for the Twin Cam. The Registers organize events of their own in addition to the main activities, and are the source of a wealth of technical and historical information about the cars. Their newsletters appear in *Safety Fast!* Information about the club and its Registers can be obtained from:

MG Car Club,
Kimber House,
PO Box 251,
Abingdon, Oxon OX14 1FF,
England.
Tel: 01235 555552
Fax: 01235 533755
E-mail: carclub@mgcars.org.uk

The MG Owners Club, based near Cambridge and founded in 1973, claims to be the world's largest one-make car club. Its monthly magazine is *Enjoying MG*. There are 40 local areas within the UK and affiliated clubs overseas.
The club has a technical advice hotline and carries a large quantity of spares. Some of the events held are very big: at a rally near Cambridge in the summer of 1996, more than 2500 MGs were lined up on the old wartime airfield at Duxford, which is now an aviation museum.
Details are:

MG Owners Club,
Octagon House,
Station Road,
Swavesey,
Cambridge CB4 1BR,
England.
Tel: 01954 231125
Fax: 01954 232106

Specialists

Both clubs mentioned previously have lists of recommended suppliers, dealers and specialised workshops.

Over the years, from the British Leyland era onwards, MG management was apathetic to the needs of MG owners and so, to cater for the demand for spares and repairs, scores of small businesses in the UK began specialising in the older cars. When, in the late 80s, the Rover Group belatedly realised there was money to be made from old MGs, it set up a subsidiary - British

Motor Heritage - which now acts as a manfacturer of spares in its own right and also licences other firms to use the MG octagon and make and sell spares under the Heritage banner.

There is now almost no part that cannot be supplied for an MGA, from a trim clip to a new bodyshell, though some Twin Cam parts, such as cylinder heads, may be hard to source.

There are too many specialists to mention them all, so the following is a selection. All can supply parts lists showing the breakdown of the components and where they fit.

One of the first businesses to specialise in parts building was the aptly-named MGA Centre at Darlington in the north of England, now known as Moss Darlington. In 1997, new bodyshells were £7000 and reconditioned versions about £5000. The Moss Group also comprises other well-known MG specialists such as M&G International, Naylor Brothers and Barry Stafford. Details from:

Europe

Customer Services Department,
Moss Europe,
Victoria Villas,
Richmond,
Surrey, TW9 2JX,
England.
Freephone 0800 281182
Fax: 0181 940 0484.

USA

Moss US,
PO Box 847,
7200 Hollister Avenue,
Goleta,
California 93117,
USA.
Tel: 0800 235 6954
Fax: 805 968 6910

Some other well-known names in the MGA world are:

Brown and Gammons Ltd.,
18, High Street,
Baldock,
Herts SG7 6AS,
England.
Tel: 01462 490049
Fax: 01462 896167.

NTG Motor Services Ltd.,
282-284 Bramford Road,
Ipswich IP1 4AY,
England.
Tel: 01473 741170
Fax: 01473 743133.

Anglo Parts (Belgium, Germany, France, Holland, Italy)
Brusselssteenweg 245,
B-2800 Mechelen,
Belgium.
Tel: 32 15 42 37 83
Fax: 32 15 42 34 45.

Also at:
Ernst Befortstrasse 2,
35578 Wetzlar,
Germany.
Tel: 49 6441 76081
Fax: 49 6441 781336

3 bis Rue des Ecoles,
F-59254 Ghyvelde,
France.
Tel: 328 26 61 00
Fax: 328 26 83 73
Storkstraat 3,
NL-3905 KX Veenendaal,
Holland.
Tel: 318 551334
Fax: 318 551395

Vic. S. Antonio C/4,
I-31100 reviso,
Italy.
Tel: 422 321 500
Fax: 422 321 470

Bob West.
Tel: 01977 703828
Fax 01977 798299

Peter Wood,
Tel: 01296 730310
Fax: 01296 738499

A wealth of information on MG clubs, specialists, models and dealers can be found on the Internet at http://www.mgcars.org.uk.

Bibliography

At the End of the Road	Timothy Whisler	JAI Press	1995
BMC/Leyland B-series Engine Data	Lindsay Porter	Osprey	1985
British Sports Cars in America	Jonathan Stein	Auto Quarterly	1993
Hambros 1779-1979	Bransen & Wain	Joseph	1979
Harley Earl	Stephen Bayley	Trefoil	1990
Maintaining the Breed	John Thornley	MRP	1956
Making MGs	John Price Williams	Veloce Publishing	1995
MGA	F Wilson McComb	Osprey	1983
MGA, MGB and MGC	Graham Robson	MRP	1978
MGB - The Illustrated History	Wood and Burrell	Haynes	1988
MG The Immortal T-series	Chris Harvey	OIP	1977
MGA Restoration Guide	Malcolm Green	Brooklands	1995
Original MGA	Anders Clausager	Bay View Books	1993
The Healey Story	Geoffrey Healey	Haynes	1996
The Sports Car America loved first	Richard Knudsen	Transport Bookman	
The Story behind the Octagon	Brian Moylan	Moylan	1995
The T-series MGs	Graham Robson	MRP	1981
Tiger, Alpine, Rapier	Richard Langworth	Osprey	1982
Works Wonders	Marcus Chambers	MRP	1995

Bound volumes: *Safety Fast!, Autocar, Motor, Autosport, Motor Sport, Road & Track, Motor Trend, Mechanix Illustrated, Motorsport.*

INDEX

Dear Reader,
We hope you enjoyed this Veloce Publishing production. If you have ideas for books on MG, or other marques, please write and tell us.
Meantime, Happy Motoring!